WHAT HAS
NATURE
EVER DONE
FOR US?

HOW MONEY REALLY DOES GROW ON TREES

TONY JUNIPER

Foreword by HRH The Prince of Wales

SYNERGETICPRESS
expanding human knowledge

First published in the UK in January 2013 by:
Profile Books
3A Exmouth House
Pine Street, Exmouth Market
London, EC1R OJH

Published by Synergetic Press
1 Bluebird Court, Santa Fe, NM 87508

Library of Congress Cataloging-in-Publication Data

Juniper, Tony.
  What has nature ever done for us? : how money really does grow on trees /
Tony Juniper ; foreword by HRH The Prince of Wales.
    pages cm
  Includes index.
  ISBN 978-0-907791-48-5 (hardcover) – ISBN 978-0-907791-47-8 – ISBN 0-907791-47-6
1. Environmental responsibility. 2. Environmentalism – Philosophy. 3. Sustainability – Philosophy.
4. Nature – Effect of human beings on. I. Title.
  GE195.7.J86 2013
  304.2--dc23
                                2013013744
-
Cover design by Ghost Design
Book design by John Cole
Typography styling suggestions by Awake Media
Editor, N. American edition, Linda Sperling

Printed by Friesens Printing, Canada
The text of this book was printed on Rolland Enviro 100% recycled Post Consumer Waste paper.
Typeface: Perpetua and Helvetica Condensed

CONTENTS

WHAT HAS NATURE EVER DONE FOR US?

*O*NE OF THE GRAVEST *misconceptions of the modern age, and one which has concerned me for more years than I care to remember, is the presumption that Nature can be taken for granted and her needs ignored. There are some who seem to think that only when times are good should we afford the cost of nurturing the natural environment. There are plenty more, I am afraid, who see the process of protecting natural systems as the sort of cost that should be avoided altogether, simply because it actively interferes with development, job creation and economic growth.*

This prevailing attitude could not be further from the truth. Nature is, in fact, the source and very basis of our welfare and economic prosperity. For me, this is so self-evident as to seem ridiculous even to say it. But as countries struggle to meet the enormous economic challenges they face, the biggest one of all remains largely hidden from view.

As you will discover in this book, the services and countless benefits to the human economy that come from Nature have an estimated value every year of around double the global Gross Domestic Product, and yet this colossal contribution to human wellbeing is hardly ever mentioned when countries consider how to create future growth. As I have long been trying to point out, this situation cannot remain the case for very much longer. We are reaching a critical turning point when humankind has to realize that people and the human economy are both embedded within Nature's systems and benevolence.

To some extent, this awareness is slowly starting to gain ground in the mainstream of our collective thinking. In part, this is the result of recent scientific studies and discoveries which are being translated into many inspiring examples of practical action. Our dependence on Nature is also slowly being reflected more confidently in those economic policies which enable people to achieve a better balance between keeping Nature's systems intact and creating economic development that results in more jobs. But if we are to deepen this commitment to Nature's needs, it is paramount that we adopt a different mind-set; one that veers away from the focus that has dominated the past half century or so. Essentially, we have to become far more joined-up in our thinking and behavior.

For example, the so-called "Green Revolution" which began in agriculture during the 1960s and quickly enabled global food production to expand and keep pace with the accelerating growth in population has also, among other things, caused the dangerous depletion of freshwater around the world, made a huge contribution to climate change, caused a massive loss of biodiversity and damaged soils worldwide. Biodiversity is absolutely crucial. You cannot simplify Nature's system and expect it to carry on operating in the way it did before. There is nothing in Nature's elaborate system which is not necessary, so to take one participant out of the dance leads to the dance breaking down and, sooner or later, this will have a serious impact on the state of human health. This is why these costs have to be taken into account if we are to see what we do in its proper context, and then an approach to food production that avoids these disastrous side effects has to take its place, otherwise we are lost. It is far too easy to believe what we see at first glance – that is, that there are huge economic benefits if we use modern farming techniques and that no alternative which does not have efficiency and profit as its priorities can possibly replace it. But if we stand back, the picture quickly looks a lot less positive. In fact, it looks frighteningly bleak because the predominant approach is effectively cannibalizing its own future by degrading the natural systems it absolutely depends upon.

The same picture emerges if you look at the way we regard the economic benefits derived from destroying the world's tropical rainforests. The soils and minerals that lie beneath the forest and the timber that comes from the trees certainly all have tremendous market values, but what about the huge role they play in soaking up the vast quantities of carbon dioxide produced by power stations, factories, cars and planes? It is a natural service which has recently been calculated to be worth literally trillions of dollars. And remember, they are "rain" forests. Take the forests out of the equation and you very quickly affect how much rain falls from the skies – which, of course, has very serious implications for our ability to generate power and produce food. And yet, we conclude that the forests are worth more to us dead than they are alive! This is an insane example of the kind of short-termism that dominates the present economic world view which, by definition, is obviously not going to help us succeed as a species in the long term. Sooner rather than later the wheels will start to fall off.

There are a wealth of examples of how Nature sustains our civilizations and economies – from the oxygen we breathe, to the soil, water and pollinating insects that produce nearly all of our food; from the scavengers that help control disease to the oceans that replenish fish stocks. To understand what Nature does for us every single day of our lives is clearly vital if we are to maintain our welfare and develop in the future. Yet, as I say, these and other natural assets continue to be liquidated as if they are inexhaustible. What has perplexed me for so many years is why we fail to put two and two together and see how dangerous this is. It is surely not for want of good science and reliable information.

As the book suggests, it is in part to do with that ancient, instinctive human tendency to grasp the short-term solution because, as hunter-gatherers, this was once necessary in order to stay alive. It is also perhaps to do with the seemingly impossible task of finding consensus on the kinds of national and international laws and policies that protect Nature, especially when the task depends upon a multilateral or global process. Some of the reasons are to be found on a much deeper level of human experience where there now abounds a disturbing lack of a sense of the sacred. This is very important. If nothing is sacred, most of all Nature, then we create the potential for the perfect kind of storm, to which it will be virtually impossible to adapt, let alone mitigate. This is why I was so pleased to see Tony Juniper's new book as, for me, it hits the nail firmly on the head when it explores how our economic system is so disastrously misaligned with the realities that enable it to exist in the first place.

Not only does it provide readers with a clear and compelling explanation as to what Nature does for us, it also offers some very strong examples of how that misalignment can be rectified – and that includes ways in which Nature's value can be harnessed even within our existing economic approach.

It describes simple things, like planting trees in city centers which would help to cool the air while giving city dwellers that contact with Nature which has such immediate psychological benefits. Thus, they would improve wellbeing and reduce the need for expensive air conditioning. On a larger scale, it also describes radical schemes like the one in New York, where the city has been given a modern water treatment system that relies upon water-friendly farming and good forestry practice. This is no small scheme and it depends upon the integrated cooperation of many thousands of stakeholders. The result of such joined-up thinking is the biggest unfiltered public water supply system in the United States, one that initially saved the city some eight billion dollars and has since dramatically slowed down the rise in consumers' water bills. They have gone up by just nine percent whereas had the city installed conventional treatment systems, that figure would now be nearer one hundred percent.

On a larger scale still, the book explains how some countries have begun to integrate natural values into their national accounts. One of the pioneers is the Central American country of Costa Rica which has taken a much more integrated view of how Nature and the economy interact, seeing them as two sides of the same coin. As a result, since the 1980s, not only has Costa Rica more than doubled its forest cover, it has also doubled the per capita income of its citizens. Dramatic examples like this should encourage us to see the tremendous opportunities there are in approaching things in a much more joined-up way. All it needs is the inspiration and unlimited capacity of the human imagination to do so.

One very positive development I have been greatly moved by in recent years, and towards which I hope I have made some small contribution via the activities and projects I have initiated, is the increasingly prominent discussion about what is known in the jargon as "natural capital." This idea defines Nature as, among other things, a set of economic assets which, if managed well, can produce dividends that flow from those assets indefinitely. This is not what generally happens at the moment. Assets such as soils and forests are often simply liquidated as if they do not need to be maintained or replenished, and it surely does not require a financial expert to point out that this is the fastest way to bankruptcy!

This shift towards seeing Nature as the provider of a set of economically vital services, rather than resources that can be used up to fuel economic growth is, for me, one of the most important conceptual shifts in history. I am pleased to say that the shift is already underway, but it needs to go much further and happen much faster. I am not so naïve as to imagine this is an easy transition to achieve, especially in such economically challenging times, but perhaps our very fraught economic circumstances at the moment offer exactly the right moment for the world to force this new attitude to break through into the mainstream.

HRH The Prince of Wales

Biosphere 2

PROLOGUE
SEALED WORLD

100: Percent of human support systems dependent on Nature

1: Number of known planets capable of supporting human needs

2 to 4 Billion: Additional people dependent on Nature in 2050

WHAT HAS NATURE EVER DONE FOR US? Vultures — and, to be specific, Indian vultures — provide an example. These birds are today virtually extinct across the subcontinent, a fact that has been barely reported in the West, and yet has had huge implications. For when India's vultures were almost gone, it became apparent that they had been supporting the wellbeing of hundreds of millions of people. The reason was simple.

For centuries, India's vultures performed an essential cleaning function, eating the flesh of the many dead animals that littered the countryside. A hungry flock would clean up the carcass of a dead cow in a matter of minutes, leaving only bones. So when the vultures disappeared, and the putrefying fly-ridden corpses were left to rot under the hot Sun, the effects were disastrous and wide-ranging.

The Indian vultures had been inadvertently killed off by anti-inflammatory drugs injected in cattle and buffalo. When these farm animals died, residues left in their carcasses were ingested by vultures – and it proved lethal to them. This soon became a problem, not least because India's forty million or so vultures were between them eating about 12 million tonnes* of flesh each year. With no vultures to clean up, there was an explosion in the population of wild dogs, which had more food. More dogs led to more dog bites, and that caused more rabies infections among people. The disease killed tens of thousands, in the process costing the Indian economy a figure estimated in excess of $30 billion.

The vultures are just one among thousands of examples of natural services that are (or were) provided for free by Nature, and which are being removed to our cost. That cost is now the subject of a new branch of economics, whereby researchers are beginning to put financial values on Nature. The hope is that through knowing more about the value of Nature it will be possible to create the tools needed to reflect that value in economic transactions. Should this happen on a sufficiently large scale, then the impacts could be profound, for the numbers being generated are huge – in many cases dwarfing the value of more traditionally quantifiable economic activities.

Natural services are beginning to attract the attention of not only academic economists and ecologists, but also governments, companies and international agencies. And that is what this book is all about – an explanation of what Nature does for us, why it is so important, and what we can do to ensure Nature keeps on doing it.

Some of the material on these subjects is quite technical and buried in academic journals and reports. In an attempt to present

---

*Used throughout this book, tonnes are 1,000 kilograms (equivalent to 2,304.6 pounds).

an uncluttered narrative I have not cited references in the book but instead compiled a compendium of material at my website – www.tonyjuniper.com. This has the added advantage of taking interested readers directly to much of the source material via links to web pages.

This vast and rapidly accumulating body of research I believe signals an emerging new era of debate. For while much of the environmental discussion in recent years has been concerned with climate change, carbon emissions and how to cut them, a new wave of attention is breaking, focused on what Nature does for us (and finding ways to keep it doing what it does).

From the coral reefs that protect many coasts to the pollinating insects that help enable much of our food to grow, awareness and attention are switching to the economic value of Nature, and crucially how to protect that value.

Before embarking on the journey to explore how these values are essential for our continued welfare and development, I'd like to start by getting some measure on how Nature works, and what it takes to replicate its functions. So our first port of call is to a remarkable experiment – one that might have more relevance to the future than was appreciated even by its visionary founders.

Biosphere 2

During the early 1980s, in the shadow of the Santa Catalina Mountains of southern Arizona, plans were laid for a remarkable and unique experiment that would shed light on how our planet sustains life: the creation of a self-contained biosphere.

This ambitious scheme eventually came to fruition a decade later, when for two years a group of eight people became the first in history

to live in a manmade biosphere. It was a project that threw into perspective just how complex, elaborate and linked is our own natural
biosphere – and just what it would take if we had to try and replicate
or recreate it.

The Earth's biosphere is basically the sum of all the different
living systems and the relationships they have with each other and
with the non-living parts of our world, such as the water, air and
rocks that enable them to function. It is the self-regulating zone of
life, shielded from the icy vacuum of space by the atmosphere on
which it depends.

For the first manmade creation of a biosphere, the terrain chosen
was that part of the southwestern USA where the giant saguaro
cacti so reminiscent of classic Western films grow. It is a remarkable
environment. From hot flower-rich bushlands, the mountains rise
to above 2,700 meters, where snow fields accumulate in winter,
feeding streams and pools as they melt in summer, Nature abounds.
On the mountain slopes, where it is cool and wet, there are lush
forests of oaks. Further up still are stands of Ponderosa Pines. The
unique combination of conditions sustains an impressive diversity
of animals and plants. Orange-crowned warblers, broad-tailed hummingbirds and cordilleran flycatchers inhabit the oak forest. At
higher levels, pygmy nuthatches and northern ravens are found.

The Santa Catalina range lies just to the north of Tucson, a teeming city of over half a million souls, with grids of traffic-packed
streets, separated by blocks of air-conditioned buildings. These two
worlds – one set out on right angles of asphalt, concrete, steel and
glass, the other a complex web of cycles, patterns, loops, feedbacks
and flows – seem utterly distinct, yet are in fact very much connected. Both systems – one of forests and deserts, the other of roads,
buildings, homes and shops – are contained inside the same biosphere. All the food, water, fuel and raw materials needed to keep

the city's vibrant economy humming along are derived from the biosphere, and the nonliving systems that interact with it.

The Biosphere complex – named *Biosphere 2* (Biosphere 1 being the Earth's) – stands about an hour north of Tucson, in a quiet and remote area in the foothills on the far side of the mountains. It looks like a vast greenhouse, made from glass and steel, with a rectangular area attached to six half-cylinder shaped buildings, over an area the size of two and a half football fields. A pair of large white domed structures flanks the main building, while a scatter of high tech facilities nearby house research apparatus, power and cooling plants and student residences. It is an impressive sight, conjuring in the mind's eye a twenty-first-century Moon base as seen from the perspective of a 1970s science fiction film director.

Built between 1987 and 1991, *Biosphere 2* was constructed to study the complex web of relationships and interactions that sustain the Earth's life systems, while at the same time supporting eight humans. Its central characteristic was that it would be completely cut off from the rest of the world.

During construction over 6,000 glass panels were laid on a steel space frame. The floor was made of concrete, but to ensure a tight seal, this was covered with corrosion-resistant stainless steel. It was totally airtight, far more so than the space-training facilities at the Kennedy Space Center, and thirty times more tightly sealed than the Space Shuttles then being sent outside the Earth's atmosphere by NASA. It set records as the most tightly sealed large-scale system ever constructed.

To enable the system to stabilize air pressure, special variable volume chambers called 'lungs' were developed. These were part of the closed system and comprised underground cave-like structures connected to large rubber diaphragms. Air moved into or out of each chamber from or to the biosphere structure as the membranes

5

expanded and contracted, gently rising and falling to keep the air pressure inside *Biosphere 2* in perfect equilibrium with that outside. This aspect of the complex would prevent the sealed structure from either exploding or imploding as a result of the pressure changes caused by the daily cooling and heating of the system as the sun rose and set.

Below ground was laid the technical infrastructure of winter heating and summer cooling pipes. Electrical power was supplied from an on-site natural gas generator via airtight power cable connections.

The idea of creating a fully sealed biosphere was the dream of John Allen. He was interested in long-distance space travel and whether it would be possible to maintain a biosphere that could sustain people in isolation for years at a time. He was also motivated by better understanding how life systems work here on Earth. He had spent decades thinking about biospheres and how they worked, and by 1984 had completed the concept for *Biosphere 2*. He was 54 and in that year founded a company called Space Biospheres Ventures, to set about the gargantuan task of construction.

Allen's focus was somewhat different from most mainstream scientists. Typically, biological and ecological research is devoted to better understanding the individual parts of systems, whether they are genes, species or even ecosystems. Allen wanted to know how the whole thing worked. There was a name for his relatively new branch of science – biospherics: the study of biospheres. It went beyond ecology, to a level where the functioning of all ecosystems together is the subject under investigation.

For this purpose, he was less interested in materials and things, and more interested in relationships – the interactions that enabled a self-sustaining biosphere to function. An additional research aim was to look at biospheres in relation to other systems and to find out how to best achieve harmonization between cultural, technological and ecological systems.

This was a project of vast ambition but Allen was equipped with the diverse range of interests and experience to make it possible. A traveler, veteran of the Korean War and a volunteer at a mountain medical center during the Vietnam War, he was also an actor, writer and poet. He was a businessman and had been awarded an MBA from Harvard. His work had taken him to all parts of the world, and on his travels he had been inspired by the diversity of living systems – from deserts to the open ocean and from the rainforests to the fields and farms of Tuscany. He was also an engineer and scientist and well versed in the technical challenges of building and maintaining a fully closed system.

He had been influenced by many thinkers, including the Russian scientist Vladimir Vernadsky, who during the late nineteenth and early twentieth century had made great advances in understanding biospherics at a planetary level. Far ahead of his time, Vernadsky was hardly heard of in the West, in part because little of his work had been translated into English. Allen visited Russia to find out more and to learn of experiments undertaken there as part of the Soviet space program. One study, called Bios 3, was taken forward during the 1970s and 1980s at the Institute of Biophysics in Krasnoyarsk in Siberia. In this program two or three people had been kept healthy in a closed system for six months. They had breathed recycled air, drank recycled water and produced about half their food inside a sealed unit.

The Russians had hundreds of doctors looking at a great mass of health data collected from the cosmonauts who took part in the study. The Bios 3 scientists, as well as those at the main Russian space research facility in Moscow, made their data available and sent researchers to work with the *Biosphere 2* team. This input proved invaluable for Allen and his team in showing how it would be possible to keep the eight scientists healthy and safe inside his complex. Some had predicted that bacterial and fungal infections and trace gas accumulation would soon damage the wellbeing of humans living in a closed system, but the Russian data suggested otherwise.

To inspire the design for *Biosphere 2*, Allen took his construction team to major architectural sites around the world. They visited Chartres Cathedral, the Roman aqueduct at Pont du Gard near Nîmes and walked the silent lines of ancient stones at Carnac in France. They studied the Temple of Heaven in Beijing. They went to the Taj Mahal in India and the Pantheon in Rome; to Uxmal in the Yucatán of Mexico, and to the Inca city of Machu Picchu high in the Andes of Peru. They searched for the designs that would best underline the purpose of the life system they planned to construct.

When I met up with Allen, he was 82, but still retained all his passion for *Biosphere 2*. Wearing a well-worn brown leather aviator's jacket, his blue eyes dart and sparkle as he told me its story. "It was initially a Russian and American joint venture," he began. "It was the time of the Cold War and cooperation with the Russians was only made possible by an agreement between Presidents Ford and Brezhnev that made an exception on certain aspects of space research. We signed a deal with the Russians at the Royal Society, London. It was organized by Keith Runcorn, the man who first set out the mechanics of continental drift."

Allen and his team expended a great deal of effort in seeking out the correct site for such an ambitious venture. It would need to be accessible, at the correct latitude and with sufficient sunlight to enable the system to work. After a long search the team settled for Arizona. "We bought a ranch near the Santa Catalina Mountains. It was an old Motorola research center," he recalls.

Having acquired a suitable location, Space Biospheres Ventures set about the vast design challenge. This went far beyond architectural questions. Temperature had to be maintained within specific limits, while all repairs to apparatus had to be done in a machine shop inside the sealed complex. No spare parts would be available from outside once the system was sealed. The glass structure had to be strong enough to resist storms, hail and tornadoes, but not cut down the sunlight that would be the source of all the productivity that would sustain life inside – including the people.

Much of their attention was devoted to landscape design, in order to make the most of both light and water, and also to ensure the environment inside *Biosphere 2* sustained the human spirit. From an outcrop of limestone rocks inside the complex were views of the tower which rose from the center of the roof of the main structure. Looking the other way, the occupants could enjoy haunting views of the northern Sonoran Desert. In quiet moments they might imagine echoes of the Apache Wars which sparked off there in 1851, ending only in 1886 with the surrender and capture of Geronimo.

Inside this remarkable facility seven biomes were constructed. Biomes are the building blocks of a biosphere – the largest unit of the Earth system, short of the entire planet. Allen is very clear as to why this was the correct level for planning the complex. "The biome was the key unit. Ecosystems are way down, three levels down in fact. From the biosphere, to biomes and bioregions, you come to ecosystems next. Ecosystems can change, and quickly. We need to see ecosystems as part of the wider system. Ecosystems are very often transient features of the landscape. *Biosphere 2* was a model for lifting up our analysis, to see the bigger picture." With this objective in mind, models of five world biomes were planned, based on rainforest, coral reef, mangrove wetlands, desert and savannah. Two other biomes set out to replicate manmade systems – agricultural landscapes and an urban area. The farming biome was developed first, then the wilderness biomes, and finally *Biosphere 2's* analogue of a city. Once the broad design of the complex was complete, Allen handed over leadership of the project to his trusted colleague Margaret Augustine. Allen wished to concentrate on science and engineering matters and to become more involved with the detail of how the system would work. He was not short of challenges, not least in relation to how the biomes would be built up.

Teams of experts then set about detailed design of the biomes, choosing the species and the types of ecosystems that would be included. By far the most complex and challenging to design were the agricultural, rainforest and coral reef biomes.

Abigail Alling, a marine biologist and whale specialist, was in charge of the coral reef, ocean and mangrove systems. She found particular inspiration in Caribbean marine ecosystems, where her team studied corals in the reef lying off the coast of Belize, though the living corals were collected at Akumal in the Yucatán of Mexico, due to its proximity to Arizona. The distance was a major factor, as the coral would need to be transported quickly, complete with life support systems to prevent it from dying, and an ocean system to which it would be introduced up and running ready to take it. The Mexicans organized a special police escort so as to speed the trucks along as quickly as they could. The coral arrived safely and was introduced to the waiting marine system. The mangrove trees and other wetland plants were collected in Florida and also trucked by road to Arizona.

Sir Ghillean Prance, then at the New York Botanical Gardens and later director of the Royal Botanic Gardens in Kew, was a key adviser in the design of the rainforest system and was aided by leading Harvard rainforest ecologist Richard Evans Schultes, who is widely seen as the originator of ethnobotany – the study of the relationships between plants and human societies. Prance came up with the design for a cloud rainforest inspired by Arthur Conan Doyle's *The Lost World*. He was also assisted in this endeavor by the government of Guyana, which arranged the collection of suitable plants.

Lowland riverine forest of the kind that floods each year in the Amazon Basin was also included in the rainforest biome. Water would be the life-blood of *Biosphere 2,* and the rainforests at the heart of the system would be vital in ensuring its continuing circulation. Reflecting on the challenges that were encountered in creating the biomes, Allen told me that "Rainforests are the most complex – probably by orders of magnitude more than some others. Prance and Schultes did a great job. They looked at the rainforest as a total system, including the Indians."

Linda Leigh, a professional range ecologist, led work on the terrestrial wilderness biomes, including the desert, which was based on the fog deserts of Baja, California. She also assisted with the development of the savannah, which was in part inspired by Allen's travels in the 1960s across East Africa.

Biosphere 2 would contain a total of about 3,800 different kinds of animals and plants. In order to maximize the ability of different bits of the system to continue functioning in the artificial conditions, an approach called "species packing" was adopted. This basically set out to ensure that if one kind of animal or plant were lost, another that performed a similar function in a particular ecosystem, such as part of the ocean or rainforest, was there to continue its work – for example, as a predator, or food for another species. A high level of extinction was expected in this microcosm and the strategy to maximize diversity from the start would, it was believed, lead to more resilience and long-term stability.

The agricultural system was designed by the Environmental Research Laboratory of the University of Arizona, with consultation from the Institute of Ecotechnics, and managed inside by Sally Silverstone and Jane Poynter. Sally grew up in London but had worked in Kenya and India with farmers in programs to boost local food security. She had also worked with the Institute of Ecotechnics on a sustainable forestry project in a rainforest in Puerto Rico. Jane was also British, with an agricultural background including experience of farming in harsh climates in the USA and Australia, where she had mastered crop propagation under very challenging climatic conditions.

The plan was to create an ecologically stable, disease resistant, tropical agriculture biome that was totally sustainable. It would need to be highly productive, supplying all nutritional and health needs, and easy to operate. It took three years before the system was sealed to develop the complex soils needed to get productivity up and the plants used to the growing seasons inside. Some 1,500 different crop cultivars were studied on site and at the University of Arizona Environmental Research Lab before choosing 150 of the most productive that they believed would best sustain the needs and health of the human crew.

Allen was very clear that the agricultural team should not copy modern farming techniques that are based on heavy chemical inputs, and instead should favor conventional methods. For health reasons, no toxic chemicals were to be used and crop pests and diseases would need to be controlled through biological means. Toxic chemicals would be unthinkable in a tightly sealed world with very fast cycling times – what was in the water would quickly be in the drinking water. With this steer, the team set out to mimic farming systems that had been used for centuries and longer in Asia, Polynesia, Europe and the Americas. By choosing several of these systems to work alongside one another, it was believed all dietary elements could be provided.

As well as a wide variety of nutritious crops – including rice, bananas, papayas, wheat, sweet potatoes, beats, peanuts, cowpea beans and vegetables – the farming system was carefully planned so as to integrate animal husbandry. Domesticated animals that were to accompany the crew on their two-year mission included four pygmy goats from the plateau region of Nigeria, thirty-five hens and three roosters (a mix of Indian jungle fowl, Japanese silky bantam, and hybrids of these), three small pigs (two sows and one boar), and tilapia fish grown in the rice and azolla pond system that originated thousands of years ago in China.

The chickens, goats and pigs would not only provide meat, eggs and milk, but also companionship. They would also recycle much of the tough plant material that humans could not digest or use in other ways. And the agricultural system was not only designed around the nutritional requirements of the crew. The landscape was carefully planned for their aesthetic needs. It was, in particular, inspired by views that Allen had seen on visits to Tuscany, where he observed that the growing of food had been beautifully incorporated into the landscape.

Allen and his team were acutely aware that at the heart of the functioning of the terrestrial biomes, and the agricultural one in particular, was one of the least known natural systems of all – the soil. The dark world of soil swarms with numberless microorganisms, worms and fungal threads that challenge even the most thoughtful ecologist with a vast complexity of relationships. These interactions are not only vital for plant growth and digesting and recycling nutrients, but are also important for keeping water and the atmosphere healthy.

A never-ending dance of give and take between soil and air helps determine atmospheric composition. The soil would be a vital aspect in determining whether or not the experiment would work, and great care was taken to get it right. Allen ordered 500,000 earthworms to be introduced. "That was straight from Charles Darwin," he says. "The worms thrived, and so did the crops."

The *Biosphere 2* team was also aware of the fundamental importance of bacteria in shaping the state of the system they were building, not least the role they would play in the soil. Bacteria are engaged in a constant process of genetic exchange, enabling them to quickly adapt to changing conditions, and to change conditions themselves. This is one of the most dynamic aspects of living systems. In order to enable full adaptability in the system as it settled down and changed over time, *Biosphere 2* was equipped with a broad set of bacteria and dozens of soil types. This would enable biological responses to changes that might take place, such as a build-up of methane gas. If there were bacteria present that were able to metabolize these gases, then that would help ensure the stability of the system. Allen believed that by starting with a wide range of bacteria it would be more likely that their orgiastic gene exchanges would help move the system towards a more steady state.

The carefully designed natural systems would of course be vital for sustaining the people, including in their mini-urban area. *Biosphere 2's* analogue of a cityscape, where the crew would live and work, was based on leading designs that would meet both the practical needs of the crew and their psychological wellbeing. All materials were carefully selected to avoid any toxic build-up once the system was closed. The Wool Bureau sponsored fitting of 100 percent wool in carpets and other fabrics so as to prevent any chemical leakage that might accompany the use of synthetic materials.

Mark Nelson, one of the eight *Biosphere 2* crew who would be sealed inside the system when it was ready, was struck by the different mind-set that had to be adopted during this design stage: "The process of designing the biosphere was incredibly interesting. Engineers and top ecologists hardly ever sit down in the same room and design something together. The engineers realized they could do brilliant and challenging engineering when it dawned on them that by contrast with their normal work their job was not only to protect humans, which is what they generally do, but to protect the life which sustains the humans. It was to be not just a stunning piece of design and engineering, but would have to sustain the microbes and everything else, if the system was to be healthy and supportive of life."

Mark Van Thillo was in charge of the vast array of machines and equipment needed to keep the vital infrastructure working. The internet was not yet in widespread use, but technology was harnessed to create what was probably the first truly paperless office. Taber MacCallum took charge of setting up a cutting-edge analytical lab to support the scientific investigations that would be undertaken. Roy Walford prepared the top-level medical facilities that would permit the critical real-time monitoring of the health of the biospherian team.

It was in the design of the living and working facilities where links between the biosphere (the world of life) and ethnosphere (the world of human culture) were most obvious. Allen set out to ensure that the world of Nature and the world of human needs and desires were designed together. Six of the biomes would need to flourish so as to support the seventh: "the city." This would need to meet human needs while avoiding toxic build-up, at the same time as all waste and water were recycled. Conditions would need to remain stable within life-friendly boundaries, and *Biosphere 2*'s equivalent of an urban area would have to play its part in promoting such an outcome.

Allen's idea was to promote co-creation, whereby both systems thrived and where one (the culture of the people) did not parasitize the other (Nature). This was in part a philosophical approach, but also a very practical one. When they were sealed in, the crew would not be able to act as if detached from their actions. They would need to pay attention to how they lived. Due to the small scale of *Biosphere 2*, compared with that of an entire planet, the cycles that supported life would work much faster, and the consequences of any changes or decisions taken by the crew would thus be seen much quicker.

For example, Allen expected that after closure the carbon cycle would work at least 1,000 times faster than outside. To quickly pick up on changes in environmental conditions and technical functioning, more than 1,000 sensors of different kinds were distributed throughout *Biosphere 2*. These would be vital both for gathering data and in ensuring the crew was kept out of danger.

To achieve the complete integration of the system so that it worked as a single unit, it was designed both from the bottom-up as well as the top-down. From top-down the biosphere would comprise seven major biomes, made up of bioregions and ecosystems. From the bottom-up the team worked with microbes, to larger organisms, to biochemical functions, to communities of animals and plants, to ecosystems, landscapes, bioregions, biomes and finally the biosphere itself.

The whole system was pieced together from the small to the large, as well as from the large to the small. And the biosphere was carefully designed so as to interact with the air and water in order for the whole thing to be self-sustaining. While modern thinking is often based on the idea of "the whole being greater than the sum of the parts," *Biosphere 2* went a big step further. The design aim in this case was for "the whole to be seen in all of the parts."

As the system was assembled, the biomes and bioregions invisibly synergized with light and microbes to create a healthy and sweet atmosphere. But could eight people live in this closed microcosm, complete with their mini-city, and come out healthy at the end of a two-year experiment? Would the farming system produce enough food? Were there unforeseen sources of trace gases that might render the experiment unviable? Many experts predicted the rapid demise and death of some of the biomes. Others questioned how long the system would last before it suffered an ecological collapse.

Here on Earth, They Left Earth

In September 1991, *Biosphere 2* was ready. A 1,900-square-meter rainforest had been created. An 850-square-meter ocean, complete with coral reef, had been prepared. A 450-square-meter mangrove wetland, a 1,300-square-meter savannah grassland and a 1,400-square-meter fog desert were all functioning well. The 2,500-square-meter agricultural area was producing the food that would nourish the team. The living quarters were ready. Apart from some electrical power, heating and cooling supplied from outside, once the doors were closed the only things that would go into *Biosphere 2* were sunlight and information.

As the Sun came up on September 26, a Crow Indian medicine man, Tibetan Geshe and Toltec Curandera sang prayers. Then Roy Walford, Jane Poynter, Taber MacCallum, Mark Nelson, Sally Silverstone, Abigail Alling, Mark Van Thillo and Linda Leigh passed into *Biosphere 2*. The airtight doors were sealed behind them, and they embarked on a two-year experiment in which they would live and work inside a closed world, one designed to be a microcosm of the Earth and its systems. They would be reliant on the little pieces of carefully selected captive Nature held inside the complex, to recycle waste, purify water, maintain the air and produce their food.

By the time of closure, *Biosphere 2* had become a major tourist attraction (Marlon Brando was among celebrity visitors) and the crew entered amid a blaze of publicity. They were a picture of health and vitality. In jumpsuits reminiscent of the super-trained Space Shuttle teams who had made history the decade before, this crew, while earthbound, was to embark on an even more unique voyage.

Like the others, Mark Nelson was prepared for the experiment through a short spell in a test module that was set up as a tiny version of *Biosphere 2*. He describes his brief taste of what was to come as

"An astounding visceral experience. It was only the size of a living room, and at any one time you could see all of the plants that were providing the air you were breathing and a big part of your food. I knew about all these things intellectually, but I wasn't expecting the extreme pleasure of being so totally connected to the system. I was part of that living system, and extraordinarily thankful it was working to provide all the services I needed. That 24 hours was one of the most astounding experiences I ever had. It really whetted my appetite."

Despite this, Nelson still found "going in was kind of a shock." He remembers how "We'd been growing food, testing systems and working together. We'd been working with hordes of scientists and technicians and suddenly when we closed that airlock there were just eight of us in this amazing living system. I went in there with the greatest possible excitement. At the time it was one great leap into the unknown. Some of the technicians were making private bets that we'd be out of there by Christmas, because of carbon dioxide levels."

The composition of the air was indeed vital. Also of fundamental importance for their day-to-day wellbeing was a healthy diet. They started with some food grown in *Biosphere 2* prior to full closure, but the agricultural system produced more than 80 percent of their food during the first year of the mission. Although their diet included a wide variety of nutritious crops, the team suffered from hunger and lost weight. In the second year, caloric intake increased and they put back on some of the weight they had lost. Despite the ups and downs in food availability, their health was excellent; indeed, certain indicators dramatically improved, such as a large drop in cholesterol levels and improved functioning of immune systems.

The success of the crew in increasing food production was in large part down to how they became more skillful in using the ultimate source of productivity in *Biosphere 2:* namely, sunlight. "Most farmers are limited by rainfall," recalls Nelson, "but for us the main limiting factor was sunfall. Any place with sun and without a plant was a waste and you soon fixed that. This didn't only help with the food situation, but also the carbon dioxide problem. We spent two years filling every possible void with plants. We grew about a tonne more food in the second year, in large part from spaces that were previously underutilized."

Despite the many strategies that were designed to promote the stability of the systems enclosed in the glass bubble, major changes were soon noted. As expected, carbon dioxide concentrations in the air fluctuated wildly. In the air outside the sealed system, the concentration of this gas was at about 370 parts per million (ppm), and although it was going up (and still is) by a couple of parts per million per year (it reached 396 ppm in 2012), the concentrations of this greenhouse gas are otherwise relatively stable.

Inside *Biosphere 2,* daily fluctuations were as much as 600 ppm, with a drop during the day as plants took up carbon dioxide, followed by a rise at night when plants naturally released it. Nelson says that "We could sit in the control room and watch the changing conditions. Every fifteen minutes or so the carbon dioxide level would be recalculated, and even when you couldn't see through a window you knew when a cloud had gone in front of the sun because the rate of CO_2 decline would go down as photosynthesis went down."

Bigger still was the seasonal fluctuation in concentration, with wintertime levels at $4,000 - 4,500$ ppm and in summer around $1,000$. The scientists worked hard to keep carbon dioxide levels more stable, through controlling irrigation water to help fast-growing plants to more quickly remove CO_2 from the air. They also harvested plant material from the savannah so as to store it and the carbon it contained.

Less expected was a temporary build-up of trace gases. Even though every care was taken to avoid materials that might release toxic gases, soon after closure something was found to be leaking them. Nelson recalls that "We detected a rise in a trace gas while doing an air analysis and the scientists outside told us that trace gas came from PVC glue or solvents. We fanned out, searching until in a dark corner of the technosphere basement someone found a small bottle of glue. The top had been cross-threaded when it had been closed, so it was releasing gases that we could detect. We resealed it and put it aside and then that gas began to decline."

More worrying from the point of view of the welfare of the crew inside was the change seen in oxygen levels. Starting with a normal atmospheric concentration for this gas of just below 21 percent, levels fell steadily, and after sixteen months reached 14.5 percent. For the biospherians this was equivalent to the levels of oxygen found at about 4,000 meters above sea level and caused headaches and fatigue, among other symptoms.

To begin with it wasn't clear why the rise in CO_2 levels was happening. One suspected source was the activity of soil microorganisms. It was thought these tiny life-forms broke down carbon-based molecules in the soil more quickly than anticipated, and when carbon atoms were liberated from the chemical bonds which had locked them into the soil's organic material, they united with the oxygen in the air to form CO_2.

This process, it was thought, might lead to an increase in the concentration of carbon dioxide while reducing levels of oxygen.

Carbon dioxide levels did not rise as much as expected, however, and experiments revealed that another factor was at work in contributing to the changes measured in the air: CO_2 was being absorbed by areas of exposed concrete inside *Biosphere 2*. Rather like calcium-rich rocks in the real world, CO_2 was being laid down as calcium carbonate in *Biosphere 2's* equivalent of geological deposits.

Looking back, Allen's view is that the "biggest engineering mistake we made was in leaving the bare concrete. We didn't paint it and so CO_2 molecules went in, carrying out oxygen. There was a drop in the oxygen. It took sixteen months, but the levels steadily dropped to the point where people were getting uncomfortable." It was decided to boost oxygen levels artificially so as to protect the health of the biospherians.

Nelson recalls that "people looking at us from the outside said it was like watching slow motion. This was in part because we were on a low calorie diet, but also because of the oxygen. It got so low that we had to inject some oxygen into one of the lungs to top it up in the facility. When it got down to about 14 percent we all trooped down there and the oxygen in the lung was around 25 percent, before it was released throughout *Biosphere 2*. People started laughing and running, and then I realized I hadn't heard a running foot in three or four months.

Being in there when the oxygen was topped up really made you appreciate what you take for granted – clean air, water and food. Without the biosphere there would be no free oxygen in the air. It's a stunning conclusion, but in there you really knew it. I went down into the lung like a 95-year-old and it was like decades of age went off the body. How many people thank the biosphere for oxygen?"

It was not only in the air that unplanned changes were observed. Some rainforest species grew quickly but suffered from weakness. The same thing happened with the woody species growing in the savannah. This was because of the lack of the winds that in Nature help produce strong and resilient wood. The fog desert began to look more like California chaparral. The mangrove did well and grew rapidly, but was different from such a system in the real world, with far less undergrowth, probably because the space frames reduced the level of sunlight.

Among the fauna, some of the vertebrate species that had been sealed in, such as the birds, ocean fish and reptiles, became extinct or greatly decreased in number. Many of the pollinating insects died out. There was a population boom in pest species, including cockroaches (although happily these undertook some pollinating functions). Several species of ant that had been deliberately introduced to help maintain the rainforest declined drastically, and were replaced by a local species that had been sealed in with no one noticing.

There were changes in the ocean. Due to the rapid growth of algae the scientists had to enter the water and remove these plants from the corals by hand to prevent the reef dying. As a result of carbon dioxide being absorbed into the sea, levels of carbonic acid built up, making the water more acidic. Calcium was added to the seawater to counter this effect as it posed a threat to the whole ocean system and had to be dealt with before serious consequences resulted, including the potential death of the corals, which would not be able to function properly in overly acidic conditions.

Despite these ecological shifts, the crew remained healthy and collected a vast body of data, and on September 26, 1993 – exactly two years after they'd entered *Biosphere 2* – they emerged. In addition to shining light on innumerable aspects of how life systems function, they had shown how a creative and healthy human culture can be sustained by a small piece of functioning Nature. The eight person crew had thrived on 2,000 square meters of farmland, a tiny bit of atmosphere and a small quantity of water. They were the first successful voyagers to another biosphere.

Mark Nelson looks back on those two years and says the experiment gave an important message: "*Biosphere 2* helped bring home the fact that humans don't come in discreet packages that are outside of a biosphere. To live inside that system and to actually feel it in your body was an amazing experience. Your body understood that you shouldn't damage the plants – it was inconceivable, and you didn't need to be reminded."

It seems that the profound messages embodied in the experiment were also noticed by a wider public. "I think for many of the people who came by to see us, it touched a nerve. We had people visiting, almost like pilgrims. It was extraordinarily touching," he says.

John Allen was delighted with the success of the mission. "They lived in there for two years and left the system more beautiful and self-sustaining than when they went in." He remains especially proud of their achievements with the agricultural biome. "Half an acre supported eight people for two years, with a half-day working that little piece of land and half the day doing other work. If you scale that up, you could support 10,000 people per square mile with a high-grade diet. Record production was achieved in *Biosphere 2*, rivaling the highest yields of modern farming."

Perhaps most important of all, *Biosphere 2* provided a powerful and practical lesson as to how our biosphere is the only life support system we have at our disposal. In addition to the demonstration that nonpolluting agriculture, without pesticides, herbicides and chemical fertilizers, could be so productive, the experiment demonstrated how it is possible to design a technosphere that was in the service of life, in harmony with the biosphere. This unique experiment also illustrated the limits to human activity, and what could be safely undertaken before life sustaining systems begin to show stress and malfunction.

Having completed their planned work, Allen and his team ceded ownership of *Biosphere 2* during a second closure experiment, which began in March 1994. Although this was planned to last until January 1995, it ended earlier. There remains some controversy as to why the second closure was terminated. Although not involved directly, Allen recalls how "everything was running well initially, and then it was shut down after just six months. No reason was given. I suspect they had problems with the soil." Later, management of the complex was taken over by Columbia University. It is today maintained as a research facility by the University of Arizona.

The pioneering Earth modeling system built at the foot of the Santa Catalina Mountains back in the 1980s was named *Biosphere 2* because Biosphere 1 was already in existence. You, I and everyone else are participants within it. It is the Earth's biosphere, and it is the only one of its kind that we know for sure exists. Biosphere 1 has some fundamental similarities with *Biosphere 2,* most obviously in that it, too, is a sealed system, and like the facility in the northern Sonora Desert, its only major input is sunlight. The rest of this book is devoted to that miraculous system: the sealed world upon which we and all known life depends. We begin our voyage through Biosphere 1 from the starting point that is so fundamental for so much of life on land – with what might be called the indispensable dirt. The soil.

CHAPTER 1

INDISPENSABLE DIRT

More Than 90 Percent: World's food that is grown in soil

One Third: Farmed soil degraded since the mid-twentieth century

5.5 Billion: Additional tonnes of carbon that could be captured each year with changed soil management

THE EAST ANGLIAN FENS were once a mosaic of wetlands, swamp forests and grasslands that stretched from north of Cambridge to where the coast of eastern England meets the North Sea at a huge shallow bay called The Wash. Today just four tiny fragments of this once extensive natural habitat remain, surrounded by some of the most intensively farmed agricultural land in Europe.

The taming of the Fens was one of the great milestones in the history of farming in England. For centuries farmers had battled Nature and water for control of the land, but in the seventeenth century a wholesale transformation from wild wetland to productive modern farmland got underway, masterminded by Dutch engineer Cornelius Vermuyden. He straightened rivers and drained vast areas with wind-powered pumps using techniques developed in Holland.

A major event in Fenland conversion came later, however: the drainage of Whittlesey Mere. Until the early 1850s, this reed-fringed expanse of water was the largest lake in lowland England. In summer it covered some 7.5 square kilometers. In winter it almost doubled in size and in places was more than 6 meters deep. It was full of fish and a magnet for birds.

While the Dutch-inspired construction of new channels and wind-powered drainage had made a huge impact on the Fens, it was the installation of the recently invented steam-powered centrifugal pump that brought about the rapid demise of this particular natural landscape. The new pump could shift 70 tonnes of water a minute, and as a result of one being put in next to Whittlesey Mere, the traditional summer sailing and fishing, and winter wildfowling and skating, came to an abrupt end. Other changes were also soon apparent.

Holme Fen lies to the south of the city of Peterborough and is one of those four places where anything resembling natural habitat remains today. In 1851 it was at the edge of Whittlesey Mere. In 1852, anticipating some dramatic implications from the ongoing drainage works nearby that was converting open water and reed-beds to wheat fields, local landowner William Wells put in place a unique and, as it turned out, very effective environmental monitoring project.

Wells asked his engineer friend John Lawrence to sink a tall iron post into the ground. He procured one that was said to have come from the iron-framed Crystal Palace that hosted the Great Exhibition in London the previous year. The post was buried into the peat and fixed at its base to timber piles that were driven into the underlying clay to make sure the post didn't move. The top of the post was at ground level.

While the post has remained the same size and stayed where it was, the land has not. Today the top of the iron column towers some 4 meters above the surface of the now desiccated fen. As the ground has contracted, some of the surrounding agricultural land now lies more than 2 meters below mean sea level, making it the lowest point in the British Isles. The neat rows of fields with long straight roads and drainage ditches today give few clues that this part of north Cambridgeshire was once the bed of a substantial lake.

The activities that led to the land literally shrinking were motivated by the financial rewards to be earned through farming the rich peaty soils. These valuable sediments had accumulated since the end of the last ice age, when sea levels had risen, blocking the exit of Fen rivers to the sea. This caused the land to become waterlogged; dead vegetation didn't completely rot and this and other organic material accumulated as deposits that today are peat. The peat took some 8,000 years to build up and was to become some of the finest farmland in the UK. And, like all farmland, its productivity and value is first and foremost determined by the character of its soil.

Out of Sight and Out of Mind

Beneath our feet, out of sight and often out of mind, soil is probably the least appreciated source of human welfare and security. More than simply a prerequisite for farming and food production, it is a profoundly complex web of interactions that enables many of the Earth's life support systems to function.

Soil is the medium through which the world of life (biosphere) meets the world of rocks (lithosphere). This is not simply reflected in the fact of plants growing out of the ground. It is a highly complex

27

subsystem of our living planet where a commingling of these two worlds takes place. As was found in *Biosphere 2*, it is also where dynamic relationships between the atmosphere and life are sustained. It is of vital importance and yet it is no more than a thin fragile skin. This is even more the case when seen against the scale of the atmosphere above and the geology below.

Soil actually seems far too small a word to describe such a complex and multifunctioning system, especially when one considers the ways in which the different components of soil can vary. The makeup of soil is hugely diverse. Its essential components are weathered rock, once-living things that are now dead, living things that are still alive, gases and water. A very approximate breakdown of the proportions of these would be rock at about 45 percent, air 25 percent, water 25 percent and organic material 5 percent. Of course, these proportions vary hugely with, for example, peaty soils comprising mainly organic matter.

Often the most important factor in the character of soil is geology. Sandy soils are gritty and formed from rocks such as limestone, quartz or granite, or from glacial, wind-blown or riverine deposits. Silty soil has finer particles and is often very fertile, although prone to erosion by water. Another major influence on the character and properties of soil is the type and amount of biological material it contains – this can be comprised of pretty much anything that has been alive, including the decaying remains of soil fauna, leaves, wood and roots.

Soil organic matter performs a range of important functions and is vital in determining how soil functions. For example, organic material in the soil can hold up to twenty times its own weight in water, and thus renders soil more resistant to the effects of drought. By storing water, soils can also reduce the risk of flooding during high rainfall periods.

The organic material in soil contains the carbon-based molecules that are the energy source that fuels what is the most important component of all – the living part. And when it comes to the complement of animals, plants and microbes living and interacting below ground, the statistics quickly get quite dizzying. For example, it is estimated that ten grams (that's about a tablespoonful) of healthy soil from an arable landscape is home to more bacteria than there are people on Earth. And these bacteria might be comprised of representatives from some 20,000 species. It is not only the numbers and diversity that are impressive – on a hectare of arable soil (that is, a patch measuring 100 meters by 100 meters) there can be a volume of bacteria equivalent to the volume of 300 sheep!

In addition to the truly tiny organisms – that is, the bacteria, protozoa and nematodes – are larger creatures, including earthworms, centipedes and various insects. This vast mass of complex living organisms undertake a number of vital jobs.

One function is decomposition. As the term suggests, this is literally the business of breaking things down to their constituent parts – and in the process liberating nutrients, thus enabling new growth. Those numberless trillions of worms, bacteria and protozoa are thus at the very base of the ecological processes that enable the productivity of living systems – at least most of those on land.

Decomposition is also so important because it is the energy source that powers the processes going on in the soil. By breaking down the carbon-based molecules in the plant and animal remains in the organic matter, the bacteria, fungi, nematodes and protozoa fuel their own growth and reproduction. Most soils are complex systems that we are only just beginning to understand. And, while the pivotal importance of earthworms has been known for some time, it is comparatively only recently that science has permitted a better understanding of the subtle roles played by other soil organisms, including fungi.

The Benefits of Dirt

Irrespective of their complexity, soils obviously deliver some fundamental benefits for humankind. Over 90 percent of our food depends on functioning soil for its continued production (the exceptions include wild sea fish and the tiny amount of produce grown hydroponically in greenhouses).

No matter how much processing, packaging and marketing goes into modern food, the production of most of it depends in the end on a vast army of nematodes, microbes and worms, many of which have not even been granted a scientific name. The next time you pick up a packet of peas or potato chips, remember who the ultimate producers of those products were – and it's not only the famous brands on the packet!

It thus seems all the more remarkable that for many people soil has taken on the cultural label of "dirt," and as such is something to be avoided, washed off or concreted over. The inconvenience of "dirt" has been the motivation for many gardens to be sealed beneath wood, tarmac or gravel and treated to liberal doses of herbicide.

And it is not only for the future food that we should pause for thought and look again at our cultural relationship with soil. Another aspect that has been on the political agenda in recent years is the role of soils in carbon cycling and storage. Organic matter, including the living components of soil such as roots and microbes, are important in this respect.

While many people have changed their light bulbs, left their car at home and debated with their friends and family the pros and cons of wind turbines, how many have thought about soil as one of the main ways we might respond to rising levels of atmospheric carbon dioxide?

Yet researchers have estimated that in the UK alone, on the order of 10 billion tonnes of carbon is stored in soils: that's more than in all of the trees in the forests of Europe. The peat-rich soils of the English uplands alone contain organic material with more carbon than all the trees in the UK and France added together. In the lowlands too, there is significant carbon held in the soil.

To put this factor of planetary stability into a wider context, it is estimated that the amount of carbon in the world's soils is more than that found in the atmosphere and all plants combined. The iron post rising from the desiccated surface of Holme Fen testifies to how that fact is subject to ongoing change. Another issue that has followed carbon on to the international agenda is the provision and efficient use of fresh water. Soils help purify water by enabling the precious liquid to enter rocky aquifers beneath them, rather than running off the land. Soils also store a lot of water themselves. These properties help supply most of the world's population with fresh water.

According to one estimate from the Environment Agency for England and Wales, a single hectare of soil has the potential to store and filter enough water for 1,000 people. Promoting the ability of soils to hold water will also help sustain food production, especially in times of low rainfall. With climate change causing more extreme weather, including droughts, the capacity of soils to store water will be an increasingly important factor for crop productivity in terms of yield and quality, and indeed food security as a whole.

Soil Stress

Farming can, however, put significant stress on soil systems. Plowing not only exposes soil to the air, but it breaks down soil structure, making the disaggregated soil particles highly vulnerable to erosion. In dry spells, when soil can lose its cohesion, the wind can whip soil from the land to form plumes of dust that can sometimes travel for thousands of miles. Other forms of soil degradation can lead to the same outcome. On March 6, 2004, a NASA satellite captured a remarkable image of a massive dust cloud heading out across the Atlantic Ocean from North Africa. It extended across about a fifth of the Earth's circumference. The air currents took a huge quantity of soil out over the sea as far as the Caribbean.

Soil turning to dust and being taken by the wind is not only a challenge for the developing world. In the 1930s, United States farmers in the midwest were confronted by the effects of soil degradation after intensive farming destroyed the protective cover of vegetation and hot dry weather turned the soil to dust. High winds in 1934 transformed an area of some 50 million acres of once productive agriculture in parts of Oklahoma and Kansas into what became famously known as the Dust Bowl.

Soil erosion is often caused by the wind, but in some regions the effects of rainfall can be more pronounced, as seen by the eroded soil in rivers turning the water the color of cocoa or tea. This is where China's vast Yellow River gets its name, as ochre-colored soil is eroded and washed toward the sea by the river's mighty transporting power. The middle and upper reaches of its catchment suffer the most serious soil erosion in the world. The degraded area here is more than one and half times the size of Britain and loses up to 1.6 billion tonnes of soil each year.

The loss of topsoil is the most serious symptom of soil degradation and is regarded as a major problem in many regions, including parts of the USA and Australia. Recent estimates suggest that each year more than 10 million hectares (25 million acres) of crop land is degraded or lost, as wind and rain erode topsoil. According to one authoritative estimate, an area of agricultural land about ten times the size of Britain has been degraded to the point where it is effectively of no use for food production. Globally, and since the mid-twentieth century, about a third of all farm soils have become degraded to some degree.

Soils generally take a long time to accumulate: a reasonable working figure is about 1 millimeter per year. Rates of soil erosion in many areas are far higher than this, and the loss of topsoil can be seen as effectively the depletion of a nonrenewable resource. In parts of the United States soil is being lost ten times faster than it is being replenished; in parts of China and India it is estimated that soil losses exceed soil formation rates by a factor of forty.

Despite these dramatic demonstrations of soil loss, in many places the effects of erosion can appear minimal, too slow year by year for the human eye to notice. But over longer periods, the loss of, or damage to, soil can lead to major change, as the metal post at Holme Fen testifies. In that case, drainage first caused the peat to shrink, and then as it was exposed to the air it became oxidized and "evaporated" when oxygen molecules united with carbon in the organic matter to form carbon dioxide. It literally turned into thin air!

And that process continues as peat soils are drained and ploughed. In the Fens of eastern England peat loss and shrinkage continue at a rate of about 1 to 2 centimeters per year. The implications for food production in this important agricultural region are clear. In some places the peat has already eroded to the point where the underlying clay is exposed. This is not only an issue for food. Recent research estimates that the exposed lowland peatlands of England could be emitting nearly 6 million tonnes of carbon dioxide per year.

Not only can soils contribute to climate change, they are also vulnerable to its impacts. While warmer temperatures and elevated carbon dioxide concentrations in the air might increase the rate of plant growth, and therefore increase yields in some places, more extreme conditions might be expected to cause an overall negative effect. This will include, for example, increased rates of erosion due to more intense rains and reduced soil moisture due to drought and changes to seasonal rainfall.

Soil is under pressure, and its ability to provide a full range of essential goods and services has been reduced, in some places quite dramatically.

How Much Soil Can We Lose?

One person who spends a lot of time thinking about soil is Professor Jane Rickson. She is the leader of the Soil Conservation and Management group within the UK's National Soil Resources Institute and a leading soil scientist. Her research lab at Cranfield University in Bedfordshire, England, has an array of facilities to enable forensic investigations into how soils work.

One experimental apparatus includes a trench full of soil with full-scale machinery running over it to simulate the causes and effects of soil compaction. Another permits the study of how different stocking densities of farm animals affect soil and how farmers might manage land to reduce erosion. There are rainfall simulators that enable any kind of precipitation to be created, from light mist to an intense tropical storm, to see what effects this can have on different soils. There is a machine to see how raindrops wobble and spin as they fall and how this affects soils when they hit the ground. These and other research tools help build more detailed knowledge of what is happening to soils, and why.

But while Professor Rickson's research work is based on a very detailed understanding of how soils work, it is her conclusions about whole soil systems that are most striking. "The soil system is more than the sum of its parts," she says. "We have excellent soil chemists, soil biologists and soil physicists who are experts on different aspects, but it is the whole system that is most vital to understand. It seems to me like a living engine. It is self-regulating and ticking over. Take one component out and the engine stops working."

"Soils are being damaged and depleted quicker than they are being replaced. Rates of loss have to be balanced with rates of formation for soils to be sustainable. The net loss of soil means that different services cannot be sustained, including food production. But how much soil can we lose before we begin to incur unacceptable and irreversible impacts? Coming up with an answer to that question is unfortunately rather complicated."

While it is difficult to give a figure as to what level of soil loss leads to what society might regard as unacceptable impacts, we do know that such a level has been reached in the past. Examples from history show how soil erosion caused social tension and even the collapse of entire civilizations. In Iceland, the Yucatán Peninsula and a number of remote Pacific Islands, the loss of soil brought about social disintegration as the ultimate source of food security was removed.

It is estimated that about 1 billion people today live in regions experiencing land degradation and declining productivity. Much of the worst damage is in China, Africa south of the equator and parts of Southeast Asia. In other words, in places undergoing rapid population increase.

During recent decades, and at the global level, the loss of farmland through erosion has in part been compensated for through more intensive farming methods and also opening of new areas to cultivation. Between 1985 and 2005, cropland and pasture expanded by about 154 million hectares, and most of this came at the expense of the tropical rainforests.

Some 24 percent of the global land surface is now cultivated. Most of this is concentrated in India, Europe and western Russia, Central

Asia, North America, eastern South America, sub-Saharan Africa, China and other parts of East Asia. Another quarter of the total land area is managed as pasture. Many of those areas where cultivation is absent are too dry, too cold or too mountainous. Opening this cultivated land and pasture has of course been achieved at the expense of natural habitats. During the last two centuries or so, we humans have converted about 70 percent of the planet's grasslands, about half the savannah biome and nearly half of the deciduous forests.

While there is still scope for expansion into new virgin land to increase the area of soil available for food production, there are very important reasons as to why those wilder places should remain as they are – and we will come to those reasons later on.

As a result of an increasingly clear squeeze between supply and demand, soil loss has become a key global issue. And as we reach the stage where it is unwise to convert more natural habitats, more and more is expected from the soils we have at our disposal. The world population is now over 7 billion and is expected to go above 9 billion by mid-century. That equates to finding the means between now and then to feed more than 200,000 extra people, every day. And the global population could go to as high as 12 billion.

At the same time as the population goes up, there will be renewed efforts to feed the 1 billion or so who don't have enough to eat now. These factors have led the UN's Food and Agriculture Organization to conclude that demand for food will increase by around 70 percent by 2050. This estimate might be too high, but pretty much all experts agree that food output will need to increase from a finite soil resource (and the oceans).

The mismatch between available soil and rising demand for food has evidently not been missed by those who are either tuned into money-making opportunities or preventing food shortages. During 2009 it is estimated that an area of land about twice the size of the

UK (some 56 million hectares) was acquired by foreign bodies, including financial institutions such as pension funds, or sovereign wealth funds managed on behalf of countries. For many of these investors, the emerging squeeze between the supply of soil and demand for food was their main motivation for risking their money.

And it is unfortunately not only food that we are expected to want more of. Demand for energy will also increase, and some of that will inevitably need to come from the land, in the form of bioenergy – including the liquid biofuels that will be grown in attempts to help fill the gap created as demand for the mineral oil from which we produce diesel and gasoline exceeds supply. Wood and other biomass will also be in increased demand to generate electricity and heat. That will add an additional demand for soil services, as will the rising demand for paper, textiles and building materials – all of which depend on soil for their production.

The coming decades will also see intensified demand for land for infrastructure and urban development. New cities and towns equivalent to between about 300 and 500 times the size of London will be needed to accommodate urban expansion by 2050. A vast area of soil will be entombed, probably permanently, in concrete and tarmac. At the edge of urban areas worldwide, productive farmland is being sealed at a rapid rate beneath houses, roads and offices.

It seems unlikely that past patterns, whereby the increased demand for the services provided by soil were met through increased supply, will be repeated in the future. One major study, the Millennium Ecosystem Assessment, recently set out how between 1960 and 2000 world food production increased by some two and a half times. This was in part made possible through improving agricultural technologies (which have in part contributed to soil degradation) and by opening new soil to cultivation. While technology will be important, on its own it will not be able to meet continued rising demand for soil productivity. With all this in mind, efforts to protect, enhance and re-store soil will need to be a big part of our forward plan.

Jane Rickson believes we need a joined-up approach that aims to promote all the essential soil services we depend upon: "Given how soil is such an incredibly complex system, perhaps the right question to ask is how best to manage this system to optimize all the services, rather than just promoting one, for example, food pro-duction. Then, we can make sure that using soil for any given service does not undermine its ability to deliver other services."

Positive Signs

Fortunately, this rather basic but vital point is now more widely understood and there are efforts underway to turn this ambition into reality. Even at the heart of the political establishment that has over decades subsidized the kinds of farming that have led to so much soil damage, there is a sense that change is required.

On a recent visit to Washington, I saw on the Metro system posters that proclaimed the efforts being made by American corn farmers to cut soil erosion. The message that came with the adver-tising boards was that soil is no marginal issue, but vital to the welfare of great countries. That this is finally getting into political campaigns is a positive sign. Perhaps Franklin D. Roosevelt's powerful assertion that "The nation that destroys its soil destroys itself" is finally being taken seriously.

This realization is in many places being backed with practical ac-tion. Reducing the need for conventional plowing through farming methods that diminish or avoid the need to break up the all-important surface layer of the soil can make a huge positive difference to soil health. Such practices are increasingly widespread in North and South

America, where several benefits have been seen – for example, in retaining water and in cutting farmers' tractor fuel costs. Long-term studies in the United States show how the loss of soil organic matter caused by endless plowing and turning over the subsoil has been halted and reversed through these kinds of "reduced tillage" methods.

The development of different crop varieties also holds the potential for the world to retain soil services while at the same time meeting the increased demands placed on it. One route is through perennial grain-producing plants. Wheat, barley and corn are derived from wild grasses that reproduce on an annual cycle. Each year they produce seeds (grain) and then die. This is the backbone of global farming, with such crops planted across two-thirds of the cultivated land.

Research is presently underway to establish the possibilities for the production of grains from grasses that grow back from the same rootstock year after year (like some of the grasses found in pastures), thereby avoiding the need for annual plowing and reseeding. There are considerable challenges, but if it is possible to breed perennial grain varieties, major advantages could be gained, in terms of increased soil carbon storage, better water retention and reduced erosion.

More radical thinkers suggest, however, that we need to go several steps further and to more closely mimic natural ecosystems in how we get the best from soils. The Agroforestry Research Trust is, as the name suggests, devoted to investigating how food can be produced from systems that look more like forests than fields. While agroforestry is quite well established as a farming system in some tropical regions, the Trust's work is devoted to understanding how such systems could work in the temperate parts of the world.

Experimental plots located near Dartington, Devon, in southwest England, are generating interesting results. Study plots were designed to be self-sustaining, and in pursuit of this goal contain a diverse range of species. The idea is that diversity will make it harder for pests and diseases to get a hold while also rendering the system more resilient in the face of weather extremes. The plots are set up so as to provide a wide variety of fruits, nuts, edible leaves, medicinal plants, poles and fibers, among other things.

Around 140 different species of tree and shrub crops have been introduced to form the "canopy" layer of tallest trees and shrubs. These range from common species like apples, pears and plums, to less common ones like azaroles, chinkapins, cornelian cherries, high-bush cranberries, honey locusts, Japanese pepper trees, medlars, mulberries, persimmons, quinces, strawberry trees and sweet chest-nuts. An area of willows that provide raw materials for baskets has been planted in a wet area of the site.

Shrubs occupy much of the space beneath trees, including the more common bush fruits (black currants and raspberries) and oth-ers such as barberries, elaeagnus, Japanese bitter oranges, Oregon grapes, plum yews and serviceberries. Many of the understory shrubs capture nitrogen and make it available to the other species.

In spring the plots are bright green, and bursting with life. Thrushes sing loudly from tall perches while bees buzz around the different blossoms. But, while much of the action appears to be going on above ground in the lush green plots, it is down below, in the soil, where perhaps the biggest difference is being made. Crucial to this are mycorrhizal fungi. These organisms help trees grow through the supply of nutrients, improved soil structure and control of disease. In return, the trees provide sugars. The fact that they can devote up to a fifth of their output to feeding the soil fungi says something about the importance of this relationship between the trees and soil organisms. Some 90 percent of land plants depend on soil fungi to grow, so this relationship is vital. And as well as aiding plant growth, the fungi are also helping to keep a lot of carbon in the soil.

Martin Crawford is the Trusts Director: "I have been surprised at how fast systems like forest gardens can become self-sustaining," he told me. "It is clear that we are only just beginning to find out what mycorrhizal fungi can do. For example, it has just been discovered that plants can send messages to each other via the fungi to warn about pest attacks, and thus to start producing defense chemicals in their leaves before a pest arrives on them. I think there are many more surprises like this to come."

He goes on to point out the many benefits that can come with an agroforestry-based approach. These range from reduced soil water loss through canopy shading, soil nutrient retention, maintenance of organic matter and the ability to recycle organic waste such as prunings and green matter into a closed loop. He says such systems can use solar energy more efficiently than crops with just a single species, and can combat pest infestations without the use of chemicals.

Importantly, given the large-scale soil damage that has taken place during recent decades, Crawford points out the potential of agroforestry in the recovery of degraded land, and how even relatively small areas can boost natural diversity in landscapes otherwise devoted to the industrial production of annual crops.

Even without radical shifts from annual crops to perennial systems, there are encouraging examples from across the world of successful efforts to help prevent the worst soil erosion and restore land that has become grossly degraded. Even in the catchment of the infamous Yellow River, erosion has been reduced through tree planting and the construction of tens of thousands of small dams on the watercourses that feed the main river. The restoration of large areas of highly degraded land on the Loess Plateau in China has also helped to reduce further loss of soil, and indeed to begin restoring what has already been lost. Terracing of hillsides, planting trees on hilltops and controlling of grazing are among the strategies that are making the difference. There is cause for optimism from recent experiences in parts of Africa too.

SCC-Vi Agroforestry is a Swedish Cooperative that has been working in the Kisumu and Kitale regions of western Kenya. Farming there had become increasingly marginal. The deforestation of hillsides had reduced the supply of water as rainfall was no longer intercepted by vegetation but rapidly flowed off the land, while much of the soil had been lost to heavy rain. Large areas of ground were bare. Deep gullies had formed, marking the effects of disastrous erosion. Farm incomes had dropped, so people had less money to spend on health and children's education, while food security had been undermined. Was there a way back from this situation?

It turns out there was, and conserving soil was the key. In this case action was taken to restore the productivity of the land not least because of concerns about rising concentrations of atmospheric carbon dioxide and the impact this is having on the global climate. Increased carbon in the soil would reduce emissions to the air, thus making a positive contribution to the global effort to keep climate change within manageable boundaries. Carbon is, however, not only a part of the climate change jigsaw; it is, as we saw before, also the fuel that powers the soil engine. As farm practices set out to increase carbon rich organic matter in the soil, other benefits soon became evident.

A combination of compost, manure, mulch and crop rotation led to a dramatic improvement in soil quality. The soil was better able to retain water and productivity increased. Farmers with small plots of land began to generate rich mixed harvests of maize, groundnuts, beans, bananas, sweet potatoes and cassava. Trees were planted on farms as well. They formed windbreaks and further reduced soil loss through root development and canopy cover protecting the soil beneath. Tree foliage also helped to feed cows, thereby supporting milk production. When mature, the trees can be used for fence posts and firewood. Trees were also replanted so as to reforest hillsides, thereby restoring water storage and supplies.

Some 60,000 farmers have participated in this project and, through a focus on soil organic matter, many have been able to contribute to a triple win: soil carbon has been increased, food output has gone up and farms are better able to withstand the now inevitable impacts of climate change – such as more intense drought.

Soil, Grazing, Dung and Carbon

When it comes to carbon, one estimate holds that at the global level it would be possible to take in and store some 5.5 billion tonnes of carbon dioxide equivalent in soil every year. That is about one sixth of global emissions, and thus presents a huge opportunity in a world where the struggle to stave off the worst effects of climate change is faltering. This is especially the case at a time of rising concern about food security.

Given the role of over-grazing in causing soil loss in many parts of the world, it is perhaps paradoxical that one of the ways in which soil quality might be improved is through intensive grazing. However, a clue as to why it might be important comes from a study of the saiga antelope in Central Asia.

The saiga is a curious-looking animal that has lived in this part of the world since the last ice age. Characterized by translucent horns and large noses, these gregarious grazers are about the size of a goat. It is not certain how many saiga there once were, but they have declined massively, and there are now only five isolated populations left. The Ustyurt saiga population lives on a plateau between Kazakhstan and Uzbekistan. It has dropped from 265,000 individuals in the late 1980s to only around 6,000 today, mostly because of

A tale of two ranches: the same land but different grazing.

poaching for horns and meat. Other grazing animals that have declined in the region include the goitered gazelle.

The loss of these animals has been long lamented by conservationists, and also by the people who graze their domestic stock in these arid lands, and with good reason. For it seems that if the vegetation that covers these lands is not eaten by grazing animals and turned into manure, then the volume of plants goes down, and desertification follows. It seems that grazing stock today is not as dense as the wild animals once were, and that this has had an impact on the state of the land to the point where some of it is turning from semi-arid desert, largely covered in shrubs and grasses which are good for grazing, to true "desert," where there is nothing for grazing animals to eat.

Some believe that the wild grazers that used to live in high numbers throughout the landscape once provided a "nutrient cycling" service and that this prevented the land from turning into desert. Without being replaced by livestock (livestock numbers declined massively after the fall of the Soviet Union), this service has potentially been lost.

Mimicking the grazing patterns of wild animals is now the focus of promising research that could point the way as to how to increase both livestock production and soil carbon benefits. In tandem with a better understanding of how wild grazing animals might help to maintain healthy soils, there is increasing acceptance that domesticated animals can be managed to similar effect. Understanding how best to do this is one of the roles of the Savory Institute.

This organization, named after veteran Zimbabwean farmer and environmentalist Allan Savory, promotes large-scale restoration of the world's grasslands through what he describes as "holistic management." This includes using livestock to restore land, which in so doing is helping to dispel the widely held view that grazing animals are, for soils, generally more a part of the problem than they are the solution. Indeed, an experiment carried out near Victoria Falls has shown how it can be possible to reverse land degradation through an increase in grazing animals. On a 2,900-hectare ranch in the Dimbangombe area the number of grazing animals was increased by 400 percent. The soil condition improved due to the way the animals were managed.

In an interview, Savory said that his methods could bring huge benefits: "By using livestock to mimic the vast herds that used to roam our planet … we are healing the soils and allowing them once more to capture and store vast amounts of both water and carbon – leading to reduced droughts and floods and beginning to seriously address climate change."

Under holistic grazing, livestock is kept in a fenced-off (or controlled) area for a maximum of three days and then not returned to the same piece of land for at least nine months. There is thus a period of intense grazing, followed by a long period of rest. This appears to mimic how grazing occurred in the distant past, when large herds of wild animals passed by an area, grazing intensively as they went, and then did not return for some time. "Because we have greatly increased livestock properly managed to mimic Nature, we now have waist-high grasses where we used to stand on bare ground. We have brought the river back to life, and it is now home to water lilies, fish and more," says Savory.

Recently this kind of grazing pattern has perhaps surprisingly emerged as one of the options that could be scaled up to help address climate change. And it is emerging as such through some unlikely routes.

Richard Branson, the colorful British entrepreneur who heads the Virgin Group, had a breakfast conversation with Al Gore during which the former US Vice President showed him his famous slide show. Branson became convinced that the world needed to adopt low carbon economic growth. Unlike other business leaders who had reached this conclusion, however, he put his effort not only into looking at how carbon emissions could be cut at source, through cleaner energy sources, but also at the question of how to remove CO_2 from the atmosphere once it had already been released.

To drum up interest in the idea, Branson launched the Earth Challenge – a one-off $25 million prize for the best idea to remove carbon dioxide from the atmosphere, cost-effectively and at scale. He hired Alan Knight, a leading proponent for more sustainable business, to run the process of finding a winner. Knight told me how Branson had an "oh-shit-I-get-it-now" moment. "I run planes, I run trains, and I'm building a spaceship!" He recalls that while it was expected that some breakthrough technology would provide the answer, the idea of biochar soon emerged as a contender. "Biochar is basically a way of taking organic material, such as wood or organic waste, turning into the equivalent of barbecue charcoal and applying that to the soil in a way that increases its carbon content, and which also improves the soil. It locks up the carbon while improving the soil, making it a dream solution. It's especially good for improving seriously damaged soil."

When in late 2011 the Earth Challenge process reached a short list of eleven finalists, three were based on biochar. A fourth to make it onto the final list was Savory's grazing method. Knight pointed out to me that, while people talk about carbon capture and storage technology, Nature does a lot of it already – and effectively for nothing.

Whereas solar panels and wind turbines have become symbols of action on climate change, it seems that soil and grazing animals could be similarly deserving, should they be managed in the correct way.

The Death Deal

Soil is a cornerstone of human welfare. Its value might be summed up with the five essential Fs we get from it – food, fuel, fodder, fiber and fresh water, plus one C, in the form of carbon capture and storage. And, while we have abused it and degraded its ability to provide the services that are so vital for our continued development, there are many proven practical steps that can ensure that these systems continue to meet the huge demands that will be placed on them in the decades ahead. This is, of course, not only about how we treat soils, but also the above-ground processes which build the carbon-based molecules which power its multitudinous life-forms.

Soil enables plant growth, and plant growth creates the organic material that gives energy to the soil organisms. The soil organisms in turn liberate the nutrients that enable plant growth. In some ways it might be said that the organisms in the soil "farm" the plants that provide their food. Soil gives life, but the deal is that in death the complement is returned. The above-ground power source that fuels the whole complex system is the Sun.

CHAPTER 2
LIFE FROM LIGHT

$3.7 Trillion: Value of carbon capture
services gained by 2030 through halving
the deforestation rate

70 Billion Euros: Annual cost of nitrogen
pollution in Europe

40 Percent: Proportion of earth's potential
land plant growth now used by humans

*MILLBROOK PROVING GROUND is a large area of testing
tracks located a few miles away from the English town of Bed-
ford. New military vehicles and the latest car models are among
the new means of transport tested there. For both security and
commercial reasons there is a need for secrecy. It has a high
wire fence and visitors are banned from taking cameras on site.*

I visited in early 2012 to speak at an event promoting a type of experimental car seen rarely even there – or indeed anywhere. It was
flat, thin and very light and made by a team of students and staff based
at the Engineering Department at the University of Cambridge.
Its shape made it look a bit like a leaf, which was quite appropriate,

considering the car was solar-powered. The top of its flat body was covered in solar panels that could generate about enough power to drive a hairdryer. But with ultra-light materials and aerodynamic design, it was sufficient to move the car and its driver along at about 100 kilometers per hour. The design was good enough for this car to have earned a respectable placing in the World Solar Challenge – a 3,000-kilometer trans-Australia road race between Darwin and Adelaide.

Looking at the solar-powered car, and all the work that had gone into making it so light as to be able to use the dispersed energy from the Sun, reminded me just how incredibly energy-dense is the diesel and gasoline that we put in most cars. So energy-rich are these liquids that just a tenth of a liter holds sufficient energy to propel a small car and its occupants from sea level up a hill the height of the Eiffel Tower.

These fuels are of course refined from crude oil, which in turn comes from geological deposits held within the Earth's crust. The oil was formed from the remains of plants and animals that lived millions of years ago. They grew and reproduced with energy derived from the Sun. When they died, they were covered with sediments that later became rock. Under immense heat and pressure, their dead remains were "cooked" and transformed into the substances that power the modern human world.

While solar-powered cars remain at the prototype stage, with bicycle-type wheels and a single occupant in a shiny suit, the fact is that nearly all of the cars on all the world's roads are in a fundamental sense also powered by sunshine.

Built With Sunlight

The Sun's vast nuclear furnace has thrust vast quantities of energy into the Solar System for more than 5 billion years – but only on one planet has the light and heat enabled complex life to develop and endure, and that is here on Earth. Green plants are the means whereby that energy is captured, transformed, stored and then transferred to animals – including us humans. The fact that we have been able to rely on so much "banked sunshine," stored in the oil and other fossil fuels, has perhaps blinded us to this basic day-to-day reality.

The molecular machinery that enables living organisms to gain energy from the Sun appears to have been first "invented" on the ancient Earth by cyanobacteria. These tiny organisms found a means to convert sunlight into chemical energy. The method they developed was passed on, with evolutionary refinements, first to simple plants and then to the hundreds of thousands of more advanced species of herbs, shrubs, trees, grasses and others that populate our modern Earth.

The process which first evolved all that time ago we know as photosynthesis – a term that literally means "to build with light." Without it only the most basic life-forms could exist – for example, bacteria which derive the energy they need to power their metabolic processes from the chemical discharges found around volcanic vents in the deep oceans. It is employed by a multitude of life-forms, ranging in size from single-celled algae through to the largest living things of any kind – the giant sequoias (redwood trees) of California. Although diverse, all the green plants rely on a marvelous process that enables them to convert inorganic molecules, carbon dioxide and water into an organic one – glucose – which in turn enables more complex compounds to be produced, with oxygen released as a by-product.

Having converted sunlight into chemical energy, plants release and use it again by breaking the bonds in glucose. They cannot store glucose, so they convert surplus output into longer chains of sugars called starch. This chemical power store enables plants to drive the metabolic processes that permit life and growth. It is the energy supply that enables the manufacture of many other molecules, such as fats, the cellulose that gives structure to plant tissues, proteins and the DNA needed for reproduction. Plants must also make the substance that enables photosynthesis – chlorophyll. This remarkable stuff is what enables sunlight to power chemical reactions that result in the manufacture of glucose. Chlorophyll is found mostly in plant leaves, in tiny structures called chloroplasts. They are housed within the rigid cellulose walls of plant cells, where they mill around jostling for position with the brightest light in which to do their solar-powered work. Chlorophyll absorbs the red and blue wavelengths of sunlight to power the reactions that enable photosynthesis. Green light is not absorbed, but reflected, which is why leaves look green. Carbon dioxide is taken in via tiny holes in the underside of leaves – and oxygen is released via the same route. Water is taken up via the roots from soil, which is also where most plants extract the nutrients including nitrogen, phosphorus and magnesium that are essential for making proteins and other vital substances. When there is drought, plants must close the holes on the leaves to prevent water loss; that means new carbon dioxide can't be taken in, photosynthesis must cease, and that is when leaves turn brown.

Thus it is the case that, from sunshine, carbon dioxide, water, chlorophyll and mineral nutrients, most of life on Earth is based. And not only does photosynthesis enable plants to grow; it is the ultimate energy source for all animal life – including us. The protein in the muscles that enable me to write this sentence, and the molecules that are the energy source for my brain, ultimately came from organic compounds made by plants. The fossil fuels that helped grow, process and distribute the food that keeps us going also came from plants – although in that case ones that lived a long time ago.

This is why ecologists often refer to the process of photosynthesis as "primary production." It is undoubtedly the most important natural process on Earth.

So fundamentally important are plants that it is rather too easy to take them for granted. Beyond the obvious function they fulfill in the production of food for animal life, they are also vital in maintaining atmospheric conditions.

Oxygen is essential for living organisms. It enables them to liberate the chemical energy stored in food. We humans breathe in air, use the oxygen to break energy-containing molecules and release carbon dioxide as a by-product.

Way, way back in the Earth's past, long before there were higher plants, never mind animals, the atmosphere was rich in carbon dioxide and methane – oxygen was a mere trace gas. As the single-celled organisms that first mastered photosynthesis began to multiply, they made organic molecules, in the process releasing oxygen as a by-product. Over hundreds of millions of years they gradually increased the concentration of this gas. By about 2.5 billion years ago, there was significant oxygen in the atmosphere.

Oxygen is the second most abundant gas in the air (after nitrogen) and the level today is at a concentration that enables advanced animal life to thrive, a situation maintained by plants. Because of this, the sky turned blue, animals exploded in diversity, sustained by a similarly rapid expansion in the diversity of plant life.

Gases released into the air at any one point take about eighteen months to be evenly distributed in the atmosphere. Winds carry and mix the air such that the breath you just exhaled will be spread evenly around the entire Earth within a year and a half. No time at all. Some of the carbon released from your first breath on the day you were born is now locked in the wood of a tropical tree, while some is held tight in plant remains in a peat bog. Other carbon

A cloud forest – helping to keep the planet cool.

atoms in that first breath, depending on how old you are, will have been in and out of plants and animals a number of times. They might have been in the body of a mouse after it ate a seed. When it died it rotted into the soil and was released again, to be taken in by a plant and incorporated into sugar that finished up in an apple. Such is the intimate commingling of the worlds of animals and plants, facilitated by our atmosphere, and of course the plants' ability to make complex substances with light.

Plants and photosynthesis also play a vital role in climatic regulation. Plants remove carbon dioxide from the air and build organic molecules with it. This is then incorporated in leaves, stems and roots before eventually becoming soil organic matter, or being stored as peat, or even being laid down for the very long-term (tens of millions of years) in seabed deposits that might one day become oil, coal or natural gas.

This aspect of photosynthesis is vital in shaping the ultimate impacts we cause on the Earth's climate. But while the world has become preoccupied with measures to reduce emissions from fossil fuels, there are other aspects to the carbon cycle that must also be at the center of how we respond.

One piece of research from the Australian organization CSIRO recently underlined why. This body looked into the number of greenhouse gases being absorbed by forests as they photosynthesize and grow, combining data relating to forest cover with climate change projections. They found that wooded areas across the Earth, from the cool boreal forests of the high latitudes to the temperate forests of the mid-latitudes and the steamy tropical rainforests, were between them taking up one third of the carbon dioxide being released as a result of fossil fuel combustion. Pep Canadell, a CSIRO scientist, described the findings as both "incredible" and "unexpected."

What was also a surprise was the estimate that deforestation was causing about a quarter of the climate-changing emissions being released each year. In other words, deforestation was causing a double whammy on the climate: clearing trees was found to be releasing massive emissions, and when the trees were gone they could no longer absorb the carbon dioxide being released by cars, power stations, factories and houses.

The economic value of the photosynthesis going on in the forests is thus vast. Even taking the low cost of carbon dioxide credits that companies must now buy via the European Emissions Trading scheme, the work being done by the forests in moderating the impact of our emissions is truly massive, worth literally trillions of euros. One 2008 review on the value of forests estimated that halving the deforestation rate by 2030 would provide carbon capture services worth around $3.7 trillion, and that enormous figure takes no account of the many other economic benefits provided by forests, such as regulation of water supplies and sustaining species diversity. The trees are, of course, doing it all for nothing.

Beyond such fundamental ecological functions, plants are also the source of building materials, drugs, landscape and inspiration. They cool cities and sustain the soil that plays such vital roles in water cycle and atmospheric regulation.

In the tropics, where the daylight varies little during the course of the year, photosynthesis can continue at a consistently high rate all year long, at least in those places where it is also wet. At the global level the annual cycle of greening is accompanied by what might be regarded as the equivalent of a planetary breath. From about November to March, the Earth exhales carbon dioxide. This is when across the huge land masses of the northern hemisphere photosynthesis slows down or comes to an end. From April to September, as plants grow and the land once again goes green, carbon dioxide levels drop down, as plants take it in to power their growth.

A time-lapse film sequence recorded from a camera aboard a high orbit satellite reveals the burst of green in spring in the temperate north and south and reveals the spread of green when seasonal rains fall in dry areas. It also shows the plumes of algae in the oceans, and how sunlight and nutrients combine to increase primary production in the seas.

Superimposed on this annual cycle is the carbon dioxide being released by the burning of fossil fuels. So with each planetary breath, each rise and fall caused by plant growth and decay, is another carbon system, one that operates over millions of years, rather than months, as is marked by the passing of the seasons. As this fossil solar energy store is burned and carbon dioxide released, it is disrupting the seasons and annual cycles that presently govern much of life on our world.

How we might re-establish an accommodation between these different carbon cycles so as to avoid disastrous levels of climatic warming is perhaps the greatest challenge before us. In the last chapter we saw the huge importance of soils in moderating carbon dioxide concentrations in the air. A large part of the answer, should we find one in time, will undoubtedly be based on photosynthesis.

But, while concerns about our impacts on the Earth's carbon cycles have in recent years been high on the international agenda, the most important historical, and indeed present, use of photosynthesis by humans has been for food production.

A Light Meal

The conditions prevailing at different latitudes and in diverse climatic zones determine the variety of farming systems that are possible.

From barley, oats and sheep in the northern temperate regions, to grapes, olives and goats in Mediterranean climates, to cocoa, oil palm and pineapples in the moist tropics, people have worked with and harnessed local conditions to maximize the benefits of photosynthesis through farming.

The Millennium Ecosystem Assessment, the exhaustive study on the state of Nature which I mentioned in the last chapter, was published in 2005 and presented what is effectively a stocktake on the natural environment. As part of this uniquely comprehensive review of human knowledge of the status of our planet's systems, the Assessment estimated which essential planetary services were in decline and which were improving.

Alongside a great deal of bad news, notably continuing forest clearance in the tropics and widespread overfishing, there were some upbeat conclusions. Two were in relation to how we have continued, over a period of decades, to increase food output from both cultivation and livestock farming per unit of land.

This is basically a reflection of our success in harnessing photosynthesis, which in turn is a verdict on our ability to use soils, nutrients, plant breeding, the use of various pesticides and water to boost those aspects of primary production most useful to us. We saw in the last chapter, however, how some of the natural services provided by soils have been impacted by steps to increase food output, and we will see later on some of the implications for water management that have emerged at the same time. And there are other consequences.

Our ongoing efforts to take more and more from primary production have been assisted by many innovations in farming. This process of co-opting an ever increasing proportion of the planet's photosynthetic potential has not only involved the conversion of natural habitats to farmed landscapes, but also the systematic removal of any life-forms which might compete with us for the products of photosynthesis. This has been achieved in part through various chemical herbicides, insecticides and fungicides, and aided by intensive selective breeding of plants. As a consequence, much of the world's agricultural land now grows industrialized monocultures of highly productive varieties of crops, with less and less photosynthesis available for other creatures or space for other plants. In its own narrow terms it has been a very successful strategy.

Our triumph is reflected in the fact that some 40 percent of potential primary production on land is now accounted for by humans. This is a huge proportion, and is seen in the crop fields, pastures and forestry plantations that sprawl across the land, and also in those areas where primary production has largely ceased because cities have replaced plants.

As we have harnessed photosynthesis, and pushed soil, crops and pasture to do more and more, we have not only intensified pest control so as to keep more of the primary production for ourselves, but also embarked on a massive program of nutrient enrichment.

Soil scientist Jane Rickson sees this as an almost inevitable consequence of how soils have been used: "As soil gets depleted or shallower, more fertilizer is needed. Less depth means less water storage, yield goes down and prices go up. Deeper soils are more tolerant and shallow soils are affected more quickly. The effects of soil degradation can be masked by more fertilizers and irrigation. These are sometimes subtle changes and hard to quantify, but they are real."

On soils where organic matter is low or depleted, and where as a result essential plant nutrients are in short supply, various methods are used to increase fertility. In recent decades the rapid increase in food output has been supported by the addition of different plant nutrients. There are seventeen of these that are essential for plant growth, but efforts to boost nitrogen and phosphorus levels in farmed soils have been especially important in keeping yields on the rise, and at a pace that has kept up with population growth. In the last fifty years or so global fertilizer use has increased about 500 percent.

When it comes to nitrogen, it might seem rather surprising that this should be such a limiting factor. After all, some 78 percent of the air we breathe is comprised of it, but in its gaseous form plants can't use it. Nitrogen in a state available to plants comprises only about a millionth of the volume that is in the atmosphere. It becomes a plant nutrient following a transformation called nitrogen fixing, which can occur in bacterial growth, among other routes. However, most nitrogen fertilizer is manufactured – a technique established in 1913 that has been crucial in maintaining industrial scale farming.

Another essential plant nutrient that can be reduced by plowing and cropping is phosphorus. This nutrient, in the form of phosphate, is vital in virtually every chemical process a plant must complete in order to grow and reproduce. In common with nitrogen, phosphorus is an abundant element, but unlike nitrogen it does not occur as a gas in the atmosphere, but is ultimately derived from rocks; to make fertilizer it has to be collected from the environment in those places where it is concentrated. Decomposing manure is one such source as are the dead remains of animals, such as their bones. Guano (bird droppings) is another and for centuries was a key source. However, these days it is mostly mined from phosphorus-rich geological deposits. The USA and China are significant producers but about half of known reserves are concentrated in Morocco, where tens of millions of tonnes are mined each year.

There are clues in these rocks as to why they contain such high concentrations of phosphorus. Rich in shark, fish and reptile bones, in these deposits there are abundant remains of long-dead animals and plants that accumulated on the bed of a fertile ancient ocean. In common with the fossil fuels, they are a legacy from light fueled life that dwelled in a remote past.

Some estimates suggest that rock phosphate supplies might become a limiting factor in food production by the 2030s. Others suggest it will be much later than this, but no one disagrees that at some stage a peak in the production of rock phosphate will be reached. As the human population continues to grow, and with it rising demand for food, this fact should be of more than passing interest.

The manufacture of nitrogen and phosphate fertilizers is one of the ways that primary production has been increased and maintained at higher and higher levels across many agricultural systems. As organic matter in soils has been depleted by cultivation and cropping, so more and more reactive nitrogen has been produced to maintain high yields. Without it there is no way that the human population would have expanded so rapidly, quadrupling during the twentieth century and reaching more than 7 billion now.

The Millennium Ecosystem Assessment estimated that, since 1960, flows of reactive nitrogen in ecosystems on land have doubled, while flows of phosphorus have tripled. This is a huge change, and has taken place in less than one human lifetime. The authors further estimated that there has been a recent acceleration in this trend, with more than half of all the synthetic nitrogen fertilizer ever used having been since 1985. This material, so vital for food production, was first manufactured just before the outbreak of the Great War. For all of human existence before that point, our needs were met without synthetic fertilizer derived from fossil fuels. Since then, we have increasingly relied on banked sunshine to support our growing demands.

It was of course a long time before 1913 that farmers worked out the importance of nutrients, even if they didn't have a name for them. Adding fertilizers to soil was an important aspect of food security well before modern agricultural technology. Manures and composts were used to replenish nutrients, and production levels were duly maintained, albeit at yields often below those of modern industrial farms.

The Nitrogen Orgy

But, while fixing nitrogen and mining phosphates have enabled us to more effectively harness primary production, other essential processes have been disrupted. For one thing, changes occur in the soil itself. Our crop plants avidly absorb the luxury nitrogen and phosphate we lavish on them, but being spoon-fed they don't need to nurture the subtle interrelationships with different soil organisms to thrive. For example, there is no need for the fungal microthreads that would normally create, store and moderate nutrient flow.

In such conditions plants can become rather like spoiled children. They grow up with a poor ability to foster and maintain relationships. Life is too easy, and the abundant supply causes essential materials to be undervalued. And that leads to waste.

Because it comes in oversupply and in a luxury form, not all of the fertilizer is taken up by the plants. Some slips past their microscopic root hairs and escapes. Dissolved in water, it runs through and over soils and off the fields and into streams, rivers, lakes and oceans. This can cause ecological havoc, for it is not only crop plants that respond so positively to an abundance of essential nutrients.

Fertilizer leakage can cause explosions in plant growth in water bodies, especially fast-growing algae. Their nutrient-powered population bursts can quickly turn water bright green, either as microscopic single-celled forms undergo population explosions, or as the filamentous kinds form dense mats of green matter. This is generally not a good thing. As the algae multiply, grow, mature and then die off, the oxygen dissolved in the water is used up as they decompose.

This can be catastrophic for animal life, as fish and other animals suffocate in "dead zones," sometimes affecting large areas of ocean. The most notorious is in the Gulf of Mexico, where the Mississippi River that drains large areas of intensive American farmland meets the sea. Nitrogen enrichment can also cause major changes in natural habitats on land. When, for example, aggressive fast-growing species of plant receive a boost, those adapted for low nutrient conditions are unable to keep up, they get swamped and then disappear.

By adding so much plant nutrient to the environment, we have inadvertently caused profound environmental change. There are also implications for human health, with nitrogen contamination of drinking water linked to problems in blood oxygen levels in infants and so-called "blue baby syndrome."

As is so often the case, our efforts to harness the benefits provided by one natural system or service has led to impacts on others. Our success in boosting primary production and taking so much of the products of photosynthesis for our exclusive use has, among many other things, led to nutrient enrichment in the environment. And recent research shows that this comes with considerable cost.

The European Nitrogen Assessment was published in April 2011. Complied by 200 experts from 21 countries and 89 different organizations, it set out to calculate the social costs across the European Union of so much extra reactive nitrogen being added to the environment. Their results should give us cause to pause before declaring our success in raising food production as an unqualified victory for humankind.

This study identified nitrogen pollution as the cause of dead zones in the Adriatic and Baltic seas and as responsible for a 10 percent loss of plant diversity over two-thirds of Europe. It suggested that the cost of nitrogen pollution at the European regional level is about double the economic value of the increased food output achieved with nitrogen fertilizers. And, in addition to the costs to the environment, it estimated that nitrogen build-up is shortening the life span of Europeans by an average of six months. Overall, the estimated annual cost of nitrogen pollution was calculated as between 70 and 320 billion euros annually. This is equivalent to between half to double that of the 2011 bailout of the Greek economy. And yet the report secured little press attention. The figures, of course, are even more sobering on a global basis. Europe is not alone in suffering the costly consequences that arise from our ongoing program to boost primary production. North America, India and China are similarly affected.

Several sources of reactive nitrogen were highlighted in the European study, but it was that coming from farming which was seen as especially serious. While controls on nitrogen emissions from power stations and vehicles have been successful in cutting pollution in different parts of the world, there is widespread agreement that far more needs to be achieved in reducing leakage from agriculture. This is not least because, as we go forward into the later decades of this century, it seems very likely that we will need to rely more on photosynthesis to meet our needs.

Energy In Real Time

Photosynthesis is the way in which the diffuse energy of the Sun has been concentrated and accumulated to power our modern economies. As we reduce our reliance on fossil fuels for reasons of climate change and inevitable scarcity of key resources, we will need to look once more toward the annual productivity of Nature, rather than the energy and resources accumulated through the work of plants tens and hundreds of millions of years ago. Wind, direct solar power, geothermal and tidal sources will all be vital in replacing fossil energy sources, but none of these can provide replacements for the liquid energy (gasoline, diesel and kerosene) that has for more than a century been so vital for transport – and especially aircraft.

With fossil fuels needing to play a reduced role in how we sustain development and welfare, plants will take center stage as we search not only for alternatives to fossil-derived energy, but also plastics, pharmaceuticals and fibers. This is a very big deal, and although we have hardly begun to understand what needs to be achieved, some pioneers are talking of a "bio-based economy" founded more on the products of present photosynthesis, and much less on fossil photo-synthesis (oil, coal and gas).

Biorefineries, which take the products of primary production to make materials that presently come from fossil fuels, would be a key part of this possible future. From crude oil we currently make a wide range of products from fuel to plastics and pharmaceuticals, and so it could be in the future with different kinds of plants as feedstock.

If you live in Europe and run a diesel vehicle, you will already be using some plant-derived biofuel blended in with the fossil equivalent. In the USA drivers are in part propelled with ethanol made from plants. Many food and drink products increasingly come in packages made from molecules that were built by plants. Sugar cane is one of the modern biological plastic factories, making the materials needed to synthesize the packages for the top brands that supply water and fizzy drinks. Corn-based yoghurt pots are in stores now, too.

A few plants, including certain varieties of algae, produce com-pounds that are chemically identical to petroleum. These organisms offer the prospect of one day making hydrocarbons from plants in ways that won't increase overall levels of carbon dioxide in the air: the plants will use this greenhouse gas as a feedstock to make fuel. Technologies heading in this direction have also emerged as contenders to win Richard Branson's Earth Challenge prize that was mentioned in the last chapter.

A shift toward a bio-based economy comes with major challenges, notably in the extent to which land and water used to make biofuels and feedstocks for plastic manufacture cannot be used to grow food. Diverting land for fuel and other materials can cause food prices to increase. But the question that logically follows from this observation is not so much about whether we should grow food or fuel, or plastics, fibers or drugs, but more one of how we are going to do all of these things on a limited amount of land. And not only that, but how are we going to harness photosynthesis while reducing our reliance on fossil inputs, at the same time as meeting the needs of a rising population – and a population that in many parts of the world also wishes to become richer and to consume more?

On top of all this, we must protect the stability of the cycles that are so vital for the proper functioning of natural systems, and the human economy that depends on them – including the water and nitrogen cycles. We must also find ways to protect and enhance the role of photosynthesis in the carbon cycle.

Bidding for Less Pollution

When it comes to controlling nutrients and reducing the impact they cause to ecosystems, there is a great deal that can be done, in the process cutting farmers' costs and improving the quality of their soils.

Pennsylvania's Conestoga River suffers from high levels of nutrient pollution. Nitrogen and phosphorus run off the fields and stimulate algal growth, which in turn depletes oxygen levels and has caused reductions in many wildlife species, including in the large area of shallow coastal waters in Chesapeake Bay. So as to protect the ability of soils to retain nutrients, and thus to maintain the productivity of the freshwater and coastal ecosystems, the local authorities looked at different ways to prevent fertilizer escaping into rivers.

Some of the soils in this area are among the most productive in the United States, so stopping farming there was not going to be a popular option. Farmers were instead invited to participate in a novel process that set out to get the biggest reduction in the amount of nutrients for the lowest cost, while preserving the ability of farmers to use the methods which best suited their circumstances. It was called a reverse auction and involved multiple sellers (farmers) competing to provide goods to a single buyer (environmental agencies), rather than a normal auction with a single seller who seeks competition between many potential buyers. Instead, farmers were invited to make bids on the basis of how much it would cost them to reduce phosphorus pollution escaping from their fields to rivers, and because it was a reverse auction bids went down as competition intensified.

There was a budget of nearly $500,000 and farmers competed for a share of the money by offering a low price for measures they could undertake to reduce the amount of phosphorus escaping from the fields and into rivers and ultimately the sea. Technicians from the Lancaster County Conservation District worked with the farmers to estimate how much they would reduce nutrient loss through different steps, including using less fertilizer in the first place, and through creating buffer zones around watercourses.

The result was not only a reduction in nutrient leakage but soil erosion was also cut back. The pilot auctions tried in Pennsylvania were deemed to have been successful and the U.S. Department of Agriculture trialed similar auctioning methods to encourage the creation of new wetlands that will help with flood control.

Working Oysters

There are also ways to deal with nutrient enrichments once they have escaped into the environment – one of which is making use of oysters.

Across many coastal areas there were once extensive areas of oyster reefs. These features, some of them hundreds of square miles in extent, were shallow water areas where the seabed was carpeted with a dense layer of oysters. These two shelled mollusks fix themselves to the sea floor, often on the empty shells of departed ancestors, sometimes comprising a layer of shell one meter thick. The crusty layer of living mollusks on top of the bed pump water through their bodies, and in the process extract microscopic food items from the water, including planktonic single-celled plants.

While we tend to think of oysters as an exotic food, less often do we see these animals as ecosystem engineers that build entire habitats. And they don't only create an environment for themselves; they provide a home for hundreds of other species that live on and between their dense layers. Among these are many kinds of fish, including the juveniles of commercially important ones. An array of invertebrates dwell among the oysters; some, such as bryozoans and barnacles, encrusted on the shells.

People of course have long exploited oyster beds, often with little thought for their future. Around New York Harbor, massive oyster reefs once existed, some 350 square miles in extent and comprising an estimated 9 billion oysters. After they were plundered for food, the animals' calcium carbonate shells were cooked in ovens to be made into the mortar that holds many of New York's older buildings together. But as the city grew, the degraded reefs were hit by sewage pollution, which killed the oysters – and some of the New Yorkers who were still eating them. Finally, during the mid-twentieth century and the industrial age, chemical pollution finished them off. A similar story can be told from many coastal areas worldwide.

Much is said about the loss of different coastal habitats, including mangrove forests, corals and sea grass beds (we will come to those later), but research by the Nature Conservancy has found that the most seriously damaged marine habitat on Earth is in fact wild oyster beds. Some 85 percent of what once existed has been destroyed, while much of what remains is degraded.

The decline of oysters of course means less shellfish – and feedstock for making cement – but other very important benefits can disappear with oyster reefs as well. These include reduced habitat for baby fish (which can grow into the bigger fish we eat) and the loss of a hard sea bottom that prevents erosion and takes energy out of waves and storm surges, thereby protecting coastal areas from flooding. There is also the matter of nitrogen removal.

As nitrogen arrives at the sea via river estuaries and causes an explosion in the population of single-celled plants, oysters help by eating them, and in the process strip nitrogen from the water. The plants are digested and then the oysters' feces ejected on to the seabed. Bacteria get to work on decomposing this, turning the nitrogen back into gas, which then harmlessly gets back into the atmosphere in its inert form.

While a little oyster might not seem up to the task of cleaning an ocean, bear in mind that an average-sized one is every day filtering up to 200 liters of water. With this kind of pumping capacity, a one hectare patch of oyster reef (assuming a low density of about fifteen average-sized oysters per square meter, and fifteen juveniles) will each day filter the equivalent of twenty Olympic-sized swimming pools. That is a lot of water, and that is why over the course of weeks and months oysters can make a big difference to the quality of coastal waters.

No wonder lots of projects are underway to restore oyster beds. Across the USA between 2001 and 2011 more than a hundred oyster reef restoration projects have been started. The main motivation has been a desire to improve water quality, while in the process improving coastal protection, fisheries and wildlife.

Around Britain, too, the potential for oyster reef restoration is considerable – and it could be linked with wind energy. In the highly polluted southern North Sea (including around the Thames Estuary) there is ongoing construction of large numbers of offshore wind power turbines. Fishing is not permitted between the tall turbine towers, meaning that large areas of undisturbed seabed are set aside at the same time as the contribution of clean renewable energy increases. The turbines are mostly in the shallow water favored by oysters, and in some cases could provide opportunities for creating more of these important and unique natural habitats. Aside from cleaning the water, the restoration of oyster beds in this part of the world would also help the recovery of highly depleted fish stocks.

Philine zu Ermgassen, an ecologist at the University of Cambridge, has been working with the Nature Conservancy to estimate the benefits humans derive from restored oyster reefs. She told me that "healthy oyster reefs are instantly recognizable, jutting like castles out of the surrounding mud. As the oysters grow into their new habitat they create yet more habitat, not only for future generations of oysters, but for the plethora of fish, shrimp and crab species which flock there for protection. The spaces between the shells create ideal nooks and crannies to escape from predators, while the shell surfaces themselves become small forests of filter feeders, such as barnacles. An oyster of the size that is typically enjoyed on the half shell with a slice of lemon is capable of filtering up to 8 liters of water an hour. Oyster restoration cannot only yield benefits through the habitat it provides, but can also directly impact and improve water quality."

Another approach to dealing with nutrient enrichment has focused on new technologies that can capture phosphorus in sewage treatment plants.

Our food contains the nutrients taken up by plants, and when we have taken energy from the molecules made through photosynthesis, we expel much of it back into the environment by way of our digestive systems and flush toilets. While traditional sources of mineral phosphorus are being depleted, there is quite a lot of it in sewage treatment plants. The more of this we can catch, the better.

Thames Water, the company that supplies and treats London's water, has installed a new system at a sewage treatment plant by the town of Slough. Its technology strips phosphorus out of the sewage water, producing a commercial fertilizer product called Crystal Green™. From this one plant about 120 tonnes of fertilizer is produced each year. There are also plants operating in Oregon, Virginia and Pennsylvania, with designs underway for further installations in California, Wisconsin, Saskatchewan and Alberta.

The technology has benefits that go beyond the production of phosphorus, producing, as James Hotchkies from the developing company Ostara explained, "fertilizer at a fraction of the carbon footprint of phosphate mined from rock thousands of kilometers from where it is needed."

The Plant-Powered Carbon Pump

A larger-scale contribution to managing the carbon cycle, in which photosynthesis can help us manage our overall impacts on climate change, has come about from a partnership between the governments of Norway and Guyana.

Guyana is one of the poorest countries in South America, and its extensive rainforests, and the soils and minerals found beneath them, could have enormous economic value. But if logging and mining happened on a large scale, there would be less carbon storage in the forests and more carbon dioxide in the atmosphere – effects that have, of course, a global dimension.

With this context in mind, and the choice between national development and the common good of the international community, in 2007, Guyana's President Jagdeo decided to invite an international partner to help to pay to keep his country's forests intact. First he wrote a letter to British Prime Minister Tony Blair. There was no positive response from that quarter, so he tried others. In 2009, the government of Norway finally stepped forward with the kind of offer he had in mind. Norway agreed to pay Guyana $250 million over five years to keep its forests intact. This figure was arrived at in part through calculating the value of the work done by the forests to capture and store carbon. By paying to keep that work going, the world would get an economic benefit reflected in the payments being made by Norway.

Kevin Hogan was an adviser to President Jagdeo and I got to know him while I was an adviser to the Prince of Wales's Rainforests Project. He told me about the pressures on Guyana to open up its forests: "Even though Guyana had one of the lowest rates of deforestation in the world, it was seeing every day how huge opportunities were making it superficially attractive, from a short-term economic standpoint, to permit massive deforestation." The pressure came from the fast-rising demand for food, fuel, minerals and metals – in other words, all the things that can be mined or harvested from land under forests – and from mining companies, logging operations and large scale agricultural interests.

Guyana's rainforests are bigger than England and Scotland combined and, if turned over to natural resources, extraction and farming could generate huge short-term revenues. It was a major step to try and put a different value on the forests, whereby the natural values they provide would be assigned a financial value, and then for this in part to be reflected in payments made by the Norwegians.

This was at the core of the agreement between Guyana and Norway. Hogan recalls how the agreement was drafted: "After a meeting between the President and Prime Minister in January 2009, the two countries embarked on a process to create a globally replicable and scalable model for how forest countries can shift on to low carbon development trajectories. By the end of the year, this process had become a formal agreement between the two countries whereby Norway pays Guyana for the climate services provided by Guyana's forests." But after a few years, what has been the impact of the deal? "The impact has been huge, and will be even greater in the years ahead," says Hogan. "Guyana is putting in place the measures to maintain 99.5 percent of its forest cover. That delivers tremendous benefits for the rest of the world."

And the money being transferred from Norway is being used for the modernization and development of Guyana: "Guyana is moving almost its entire energy sector to renewable energy. A combination of hydropower for those connected to the grid and solar energy for indigenous and other forest communities. For 11,000 families, this is the first time they are gaining access to electricity." The money is also being used to provide 90,000 low income households with laptops and access to training in IT, to create a new generation of citizens equipped for low carbon jobs. The lands of indigenous peoples are being given proper legal title, while strong systems to safeguard the forest are being put in place.

Having highlighted this positive example of what is clearly possible, it is important to say that the costs of this kind of program might be higher in other parts of the world. Research suggests that in Southeast Asia the costs of slowing down deforestation would need to be set well above the levels for Guyana. This is not to say that it is not possible to create economic incentives to keep the forests standing, not least because it would still be cheaper to do that than to invest in carbon capture and storage technology. But new approaches will be needed in mobilizing the finances needed to change patterns of forest loss.

Bio-Based Economy

While there are more and more examples of positive practice in using primary production to achieve more sustainable outcomes that can be brought to scale, we have hardly begun to build a sustainable and renewable bio-based economy. But we could do that if we can find the means to harness photosynthesis while at the same time protecting other vital services provided by the soils and aquatic systems that are so heavily impacted by our gargantuan effort to boost and harness the benefits provided by the primary production undertaken by land plants.

For a bio-based economy to emerge, one that is fully integrated with ecosystems and what Nature can provide indefinitely, we need not only soil, water, nutrients and photosynthesis, but also the vast wisdom accumulated by animals and plants derived from billions of years of evolution.

CHAPTER 3
ECO-INNOVATION

27 Percent: Heads of global companies who say loss of natural diversity could cut growth in their businesses

25–50 Percent: Proportion of $640 billion pharmaceutical market based on natural genetic diversity

Larger Than Germany: Area of forest cleared 2000–2010

THE FLINDERS RANGES lie in the Australian Outback, some 500 kilometers north of Adelaide. Here the vast flat plains of the interior are dissected by ancient crumpled rock outcrops, sharp bluffs and soaring cliffs. It is a humbling landscape. Its scale and apparent emptiness make living things feel small.

The Ranges comprise layers of reddish, grey and brown sediments, laid down over hundreds of millions of years into an ancient ocean basin. They have been crushed by the vast pressures of sediments that built up to depths of many kilometers, contorted and twisted by the tectonic forces that thrust the layers upward to create the Ranges, and then weathered and eroded by millions of years of ice, frost, wind and water. These hillsides, littered with shattered rock and boulders, tell the story of the Earth as it was more than half a billion years ago.

Brachina Gorge, in the central Ranges, is an intimate landscape by comparison with these huge vistas. The sparsely wooded sides of the gorge are steep and narrow. There are clumps of native pines and multi-stemmed eucalyptus trees known as mallees, but otherwise there is little to curb the runoff from the periodic rains. The boulder-strewn creek bed is replete with evidence of powerful floods. The deluges have piled the huge dead branches of river red gums up against the trunks of the many living ones. Some of the young trees were flattened by a wall of water that followed recent rains, but their deep roots keep them anchored to the creek bed. If they can hang on perhaps one day, they will reach the stature of the majestic old trees which are hundreds of years old and which seeded them.

In the bed of this dramatic gorge, some special fossils can be found. In a layer of rock known as the Rawnsley Quartzite are traces of animals that lived a very long time ago. I searched for them in the failing light of a midwinter afternoon. While my hands traced some faint shapes, I reminded myself that the shadowy impressions in the rock were a connection to a quite different world. First described by Australian geologist Reg Sprigg, the fossils he found near to this spot led to a new geological era being recognized – the Ediacaran.

The Ediacaran began 635 million years ago and lasted for around 90 million years, during which time the region was a shallow sea. The sediments Sprigg found, and which I was looking for, showed marine animals, and the marks they left behind suggest that they included several different creatures, including animals that appear similar to modern segmented worms, sea pens and jellyfish. All of them had soft bodies; there were no bones, shells, teeth or other hard parts, and that is why fossils from this period are all the more remarkable. But the significance of these traces comes not only from the unlikely fact that evidence of these animals are discernible after more than half a billion years of geological upheaval and erosion, but because they are the oldest known fossils of multi-celled animals – and not just of these kinds of animals, but any kind of animal. The layer of rock in the quiet and remote creek records the period in Earth's history during which life mastered the incredibly complex challenge of making animals, at least in the sea.

The clues left in the rock, some of them circular shapes between about 2 and 6 centimeters in diameter, mark this amazing watershed. Before this point, the world was populated by only simpler organisms comprising a single cell. These included photosynthetic algae and the bigger but still single-celled animals that ate them.

The Ediacaran was the last chapter in the saga of life before the Cambrian period – the time when the diversity of life literally exploded. In fossil beds laid down 545 million years ago, most famously seen in the Burgess Shale formation of the Rockies of British Columbia in Canada, a great diversity of creatures appear, including most of the basic prototypes of modern animals. Following more than 2 billion years during which the Earth supported simpler life, suddenly the diversity and complexity of animals took on new dimensions, including the shells and other hard body parts that are much more obvious from later fossils.

I often travel on the train between London and Cambridge, a journey of 91 kilometers that lasts about an hour. An imaginary journey on this route makes an interesting yardstick against which to map the relative timescales that mark the major events in Earth's history since that first explosion of life that began at the start of the Cambrian period and which is apparent in the impressions seen in Brachina Gorge.

As the train departs Kings Cross station in central London, picture it passing the sea in which the animals of the Cambrian era lived some

545 million years ago. As the station buildings are left behind, there are trilobites, shelled animals and a myriad of marine invertebrates.

Shortly after, about 540 million years ago and just 1 kilometer after departure and long before the train passes the first of the twenty-four stations that lie between here and Cambridge, the first land plants appear. Recently evolved from simpler algae, they grow under blue skies and the land begins to go green. A little further on, and still in London, and still before the first station, the train passes the time of the first fish. They appear about 510 million years ago and are the first creatures with backbones.

The train runs along to arrive at the time of the first insects. These appear about 407 million years ago, just where London's orbital M25 motorway crosses the rail line. From creatures that resemble modern lungfish evolve the first four-legged land animals. These primitive amphibians haul on to the land during the mid-Devonian period, some 397 million years ago – near Welham Green station. By the time we travel a little further north to arrive at Hatfield, still close to London and at 377 million years ago, the early fish we saw in London have died out, to be replaced by a multitude of more modern and highly evolved versions.

Just over halfway and at some 230 million years, between Hitchin and Letchworth stations, the first dinosaurs are seen. A little later, and at Letchworth the first furry creatures appear in the form of early mammals. They are followed a while later, in the Jurassic period, just before Ashwell and Morden station, by the first feathered beings – the birds.

A couple of kilometers beyond, at 140 million years ago, the first flowers are seen. A few kilometers on, past Royston, we reach 100 million years before present and the flowering plants are widespread – and so are the pollinating insects that enable them to reproduce. The world is beginning to look more like it does today and, although it's much warmer, with polar ice caps far smaller and sea levels higher, it is cooling down. Chalky deposits are accumulating on the seabed as tiny organisms with carbonate shells build up in successive layers and, as they do, taking carbon dioxide out of the atmosphere.

The train speeds north toward Cambridge. As it approaches Foxton, the last station before our final destination, an asteroid collides with the Earth. It comes down in the present-day Yucatán region on the Gulf Coast of Mexico. In the aftermath of a massive explosion, much of life on Earth is wiped out. This is the fifth time such a massive loss of species has occurred since the train left London. We have arrived at 65 million years ago, the point at which most dinosaurs disappeared.

Only one major dinosaur tribe survives the asteroid – the birds. As life recovers on the final stretch into Cambridge, it is their time, and that of the mammals. Many of the mammals still lay eggs like their reptilian ancestors – but others now give birth to live young and feed them with milk. Grazing animals are adapted for grinding up and digesting fibrous plant material. The cats, bears and dogs have evolved to prey on them. It is 50 million years ago and several other new mammal groups have appeared, including the first primitive whales – and the primates.

The early primates look rather like little squirrels, but they have thumbs that enable them to grasp and manipulate objects in a whole new way. Animals that resemble modern lemurs follow, and then, about 34 million years ago, as travelers gather their papers and bags ready for arrival in Cambridge, early monkeys and apes appear. Many different kinds come and go; some are evolutionary dead ends, while others thrive.

About 4 million years ago, less than 1 kilometer from platform one, where I will get off the train, a group of apes begins to walk on two legs. There are many theories as to why this might have been the case, but it seems most likely linked to freeing up the hands, and those thumbs, for tool use. With about 500 meters to go, the first animals in the genus *Homo* appear. This is the group of mammals to which we *Homines sapientes* belong, and beings anatomically equivalent to modern people are seen just 34 meters from the ticket barrier. We are now just 200,000 years before present, off the train and walking to the Cambridge station exit. These early humans were, until about 100,000 years ago, confined to Africa. They reached the behavioral equivalent of modern people about 50,000 years ago, just 8.5 meters from the ticket barrier.

With just one step to go, about 1 meter left, the first crops are planted and the first towns appear. As I present my ticket, the rapid rise of humans is accompanied by changes in the Earth's vegetation cover — for example, there is now less forest, while some animals begin to disappear as they are hunted to extinction, first by spear-wielding hunter-gatherer groups and then with crossbows and finally guns.

Despite the evident impact on large mammals — saber-toothed cats, mammoths, giant sloths and others — as I place my ticket into the machine to open the barrier, and arrive in the here and now, the diversity of life on Earth remains rich, and indeed more so than most points since the train left King's Cross half a billion years ago. And that diversity didn't just appear — it has accumulated over an immense period of time.

In drawing the picture of evolution tracing the relationships between different groups, biologists present a diagram that looks rather like a tree. At its base, in the trunk, are the major groups that appeared in the Cambrian period. Major boughs branch off, and subdivide to thick branches, then thinner ones, and finally to the twigs of the individual species.

The life seen at each bough and branch was dependent on the achievements of life at earlier stages. Things that didn't work disappeared, while those inventions that met needs at any particular point were kept and built upon. If there had been no amphibians, there would not have been mammals. It is at the pinnacle of this accumulated complexity that modern humans reside. The vast diversity of life that we live amid today is like a huge library of information, relationships and systems. For us humans, it is the biggest asset of all.

As the ticket barrier opens and I look even one step forward, how much longer that incredible natural diversity will remain is not certain. The rate of species loss is now about 1,000 times higher than the time before the first humans appeared. Because of the work of evolutionary biologists, I can look back over the hypothetical 91 kilometers and see the unfolding of life on Earth. But even 2 centimeters further forward – the equivalent of a little more than a century – it is unclear as to where the onward journey leads. In that tiny distance there are two possible outcomes – and one leads quickly toward a sixth mass extinction, perhaps on a scale equivalent to that which accompanied the demise of the dinosaurs.

But why should we be bothered, even if it is the path toward a sixth extinction that we take in the step beyond the ticket barrier? As I walk from the station and my primate thumbs skip with precision across the touchscreen of a smartphone, surely we have arrived at a place where the procession of natural history is no longer of importance to us. Is it not the case that our amazing intellect has taken us out of evolution – beyond Nature?

The answer to that is, no, it hasn't. The natural diversity of animals, plants and other organisms that is all around us enables the different systems that sustain life to function. More and more research confirms how life supports the conditions for life, through cycles that maintain soil fertility, climatic stability or via population checks and balances, among other things. It is an integrated system, and we are in it and dependent upon it as much as the birds and flowers.

It is little wonder, therefore, that Nature has inspired humans since earliest times, as witness the paintings found deep inside caves, where people were once drawn with dim lights to depict, with pigments made from blood, berries and soil, images of the animals they knew they depended upon. We can only guess as to the thoughts of the artists as they descended into the gloom to paint, but it seems they held insights as to the place of people in Nature. Their minds were just like yours and mine, but what a difference in perspective is expressed through the faint visual echoes of the game packed plains!

Artistic and spiritual inspiration is, however, but one aspect of how Nature has shaped our world views and wellbeing. For not only are we surrounded by an accumulation of species seen in an eye-catching array of colors, shapes and combinations, but also an immense diversity of relationships, which in turn are reflected in the structures, chemicals and cycles that hold different systems together. These relationships take many forms – for example, that between predator and prey, plant and pollinator, tree and browser, grass and grazer, parasite and host, symbiosis, commensalism and others.

This natural diversity is also an expression of practical problem solving. In the testing and constantly shifting conditions imposed by Nature, all forms of life must continuously hone their ability to survive – or die out. The vast majority of species that have ever existed have quite naturally disappeared, to be replaced by others in the evolutionary journey that we chart today through the marks, imprints and remains that have survived as fossils.

The evolutionary processes that enable new organisms to evolve as others die out are at one level based on genes. This is the chemical life code that specifies animals, plants and other life-forms. The code is written into DNA molecules that in turn comprise the combination of genes that are specific to each different life form, determining how they appear, behave and function.

It is into this genetic language that solutions are written for meeting the challenges of survival. These solutions are not only reflected in the structures and forms we can see with our eyes, but in the vast workshop of molecular tools and arsenal of biochemical weapons possessed by living things. Some of these have already been deployed in the service of humankind. Many, many more await discovery, including in the soil.

Chemicals Arms Race

As we saw earlier, in the dark world below ground dwells a vast and largely unexplored diversity of life, with the potential for practical application in meeting many of our modern-day challenges. Through an electron microscope, the Actinobacteria look rather like fungi, with long filaments that extend through the soil where they perform several vital functions, including the fixing of nitrogen and decomposition. In order to survive in their testing environment, they have developed chemical defenses that have become the basis of most modern antibiotics. One group, the Streptomyces, produce over two-thirds of the clinically useful antibiotics that are from natural origins, including Erythromycin, Neomycin, Tetracycline and Cefoxitin.

The arms race that took place for hundreds of millions of years in the soil is now replicated in hospital wards and on farms. We fight back against the assault of microbes with antibiotic substances invented in the struggle between organisms that we can't even see. And the fight back against bacterial infection will never end – and neither will the potential ability of the microbes to get past our defenses. We can and should use antibiotics more carefully, especially on farms where their overuse to prevent infection (rather than to

Chemicals in the skin of this red-eyed tree frog can block HIV infection.

treat it) has accelerated the emergence of resistant bacteria. These then invade hospitals, where some antibiotics are useless against them. And we should also recognize the vital importance of other antibiotic substances already invented by Nature, and which we haven't found yet.

Soils host other organisms that might assist in making future medical breakthroughs. For example, the natural toxins produced by some other soil bacteria are among the most effective known anti-cancer agents. And the vast unexplored life diversity in soils also has potential for commercial exploitation in other fields beyond medicine, including in formulating new industrial processes, agriculture and in the cleanup of pollution.

It is not only microorganisms that have contributed to our health and wellbeing. Horseshoe crabs are derived from an evolutionarily distant past. Looking rather like the trilobites that appeared during the Cambrian explosion, these animals' ancestry runs back some 350 million years. They are not in fact crabs in the normal sense of the term, but are more related to spiders and scorpions.

Among the survival solutions that these creatures developed on their long journey includes a straw-colored blood that is based on copper. In their marine world the crabs live in a soup of bacteria, and their blood is a carefully honed response to the ever-present threat of infection. These animals have unusually large blood cells which manufacture a clotting agent that has been found to have rather useful properties. When it comes into contact with bacterial toxins, a clotting reaction occurs. This property has been harnessed to test the sterility of drugs, vaccines and other medical applications. The crabs are caught in the sea and transported to laboratories, where some of their blood is taken. Up to 30 percent is removed, and then the crabs are released.

The blood of these primitive creatures has saved many human lives (and also those of many rabbits, which were previously used to test the sterility of batches of drugs). To date, no synthetic test has been invented to match the reliability of that derived from the blood of the horseshoe crab. If you've ever had an injection, then it is a very good thing that these animals are still around. The fact that they endured for so many millions of years, to have survived the great events that caused so many others to pass away, and to have shared with us some of the secrets held in their remarkable blood, has made routine life-enhancing interventions far safer.

Survival solutions developed by a group of animals called cone snails might also one day enhance human wellbeing in profound ways. These creatures again have a long ancestry, belonging to a group of mollusks which first appeared in the Cambrian era. They are predatory animals and live in the shallow tropical waters of mangroves and coral reefs, where they use a cocktail of toxins to paralyze their prey. They use combinations of different toxic protein molecules delivered into the body of the unfortunate creature they plan to eat via a kind of hypodermic needle in their proboscis. They constantly revise the mixture of toxic compounds so as to prevent the creatures they hunt from evolving resistance.

Cone snails are coveted and collected because of the wonderful intricate patterns on their shells, and have been a valued item of trade for hundreds of years. It is, however, the substances they evolved to immobilize their prey that could be of far more value than their paradoxically ornamental armor. Some of the toxins used by cone snails are under investigation as potential new treatments for aspects of chronic nerve pain that have proved largely resistant to most known painkillers.

A deadly poison is also under investigation for use in heart surgery. The venom of a Central American bark scorpion is being studied for the role it might play in improving the success of heart bypass operations. Trials have also recently taken place in the treatment of a type of cancer called high-grade brain glioma using the venom of the yellow Israeli scorpion. Researchers found that the poison used by these animals to paralyze their prey contains a molecule that attaches itself to the tumor cells.

Cancer treatment might also be assisted with chemical compounds found in jellyfish. Recent research shows how the luminous cells in these animals could revolutionize the way in which cancers are diagnosed. A technique using the green fluorescent proteins made by jellyfish in conjunction with a special camera can be used to find cancers at an early stage and deep inside people's bodies.

Starfish are from a group of animals with ancestry stretching as far back as that of the jellyfish – to before the Cambrian explosion. They too have innovated for survival in ways that might assist us humans. The spiny starfish that lives around the west coast of Scotland has a slimy coating that prevents things sticking to it, thereby reducing its risk of contracting disease. The slippery substance that provides this protection has been found to have properties that might inspire a new generation of anti-inflammatory medicines which could be used to treat conditions including asthma and arthritis.

When it comes to life in the oceans, recent research has revealed how a veritable treasure trove of biological material, including millions of previously unknown genes and thousands of proteins, exist in marine microorganisms. These offer huge potential in developing new medical treatments and industrial processes. Solutions invented by Nature might also contribute toward making our exit from the age of fossil fuels.

Grazing animals have come to depend on bacteria living in their gut that have evolved enzymes which break down cellulose into its constituent sugars, in the process liberating its nutritional value. The animals can't make these enzymes, but the bacteria do it for them. Harnessing these enzymes could be one stepping stone toward new biofuels made from the cellulose in plant stems and leaves, rather than grains, fruits and other plant materials that could otherwise be human foods. This in turn could help reduce the conflict that has arisen in recent years as demand for energy crops has diverted some food into fuel tanks. By copying the enzymes made by the microbes, sugars can be split out of the cellulose that has been built by photosynthesis. These can then be fermented and turned into a renewable alternative to gasoline.

One inspiration in the development of a new generation of fuels comes from elephants. These creatures break down about a tonne of fibrous cellulosic material in their gut every week. Understanding how they do so has led to new, more efficient ways of making biofuels.

As we saw in the last chapter, plants might one day be the foundation of a whole new bio-based economy. Powered by the Sun, and in part inspired by enzymes produced by microbes that dwell in the guts of herbivores, new biorefineries could one day replace the petrochemical plants that are today such a fundamental cog in our fossil-fueled civilization. Moving from stored sunshine to annual sunshine will not solve all our problems but with the need to phase out fossil fuels ever more pressing, the Earth's genetic wealth will be an essential component in making the transition to the solar age.

There is increasing interest, too, in using natural design in engineering and industry. Seeking solutions to challenges in the human world by drawing on designs developed by Nature is an increasingly mainstream discipline known as biomimicry – literally the process of mimicking life. Janine Benyus has charted the rise and scope of this new way of looking at life in her book *Biomimicry*. There are a great many examples.

Scales from a butterfly wing have inspired new paints that don't use pigment, and therefore can avoid some of the pollution involved in manufacture. The microscopic structures that enable geckos to walk on vertical walls are behind innovations for new adhesives. The lobes on the fins of humpbacked whales have inspired designs for more efficient wind turbine blades. A desert beetle is the source of a new method for collecting water from mist in arid regions. A substance used by mussels to anchor their shells on rocks is behind new materials that could provide huge benefits in dentistry and marine engineering, among other things.

The structures in the mounds of some species of termite are being used to develop super efficient buildings. The boxfish has lent its super strong body design to the motor industry, enabling stronger cars to be built with less steel. A species of deep-sea sponge known as the Venus flower basket has structures which have inspired stronger fiber-optic cables. The surface of lotus leaves causes dirt to naturally run off when it rains, thereby keeping leaves clean so that they work better –

a design solution now being used on building exteriors, so that they don't require cleaning, saving chemicals and costs.

Great potential has been found in the design of shark skin. The rough exterior of these ancient fish has been copied to make boats go faster and a leading manufacturer of competitive swimwear has used shark inspiration to design a new generation of costumes. Shark skin technology, when applied to aircraft, in paint, can increase speed and cut fuel use. "When applied to every airplane every year throughout the world, the paint could save a volume of 4.48 million tons of fuel," asserts Professor Julian Vincent, former President of the International Society of Bionic Engineering.

Biological problem solving has been harnessed in a new roof design to dramatically reduce the need for air conditioning in hot climates, thereby offering the potential to save a huge amount of energy – an especially important innovation in a world that is warming. There is a species of tiger snake that in cooler climates might offer new ideas on how to do the opposite – to make heating systems more effective. If the snakes can do it biologically, then the chances are we can harness the same method with technology, so long as we know how they do it.

These particular tiger snakes live at the other end of the geological formation in which the Flinders Ranges are located, at Kangaroo Island. Much of Australia's rich native wildlife is still found here, and is the subject of intense scrutiny by Peggy Rismiller, who runs the Pelican Lagoon Research Station. The climate on this Southern Ocean island can be raw and wild. To the south lies open sea until Antarctica some 4,500 kilometers distant. Storms blow up from the south and bring testing raw weather to these shores, especially for the several species of native reptiles. Unable to warm their bodies themselves (by contrast to us mammals, which do it so well), they must find other solutions to the job of thermal control.

The snakes are venomous and quite secretive, and therefore quite hard to study, but Rismiller has spent a great deal of time seeking them out so as to better understand their lives and ecology. They come in seven different color variations, she tells me, "but no matter what color they are they can flatten the first third of their body to expose black skin between the scales. When they do this they can heat themselves up even in quite weak sunshine. As a result they can fill a lot of different niches all the year round. But how is the tiger snake such an effective solar collector? If we could understand this we could use the methods developed by the snake to make solar heating technology more efficient, to collect heat during the day and then use it at night."

There is also vast scope for improving structural materials through a better understanding of how Nature approaches the challenges of survival. Nacre, or "mother of pearl," is the coating found inside the shells of oysters and some other bivalve mollusks. These animals utilize the same material found in chalk to create quite different properties to the powdery and soft consistency normally associated with the sticks used to write on blackboards. The very different character seen in mother of pearl is achieved through an arrangement of platelets that are corrugated and stacked in a very precise way such that if pressure is applied no single crack appears.

A close look at wood also reveals important tips on structural design. The timber of hardwood trees has tiny holes, which collapse when a side load is applied. This is what makes them much stronger than most soft woods derived from coniferous trees. Solutions inspired by this natural innovation could help to save materials through a focus on how structures are designed, rather than simply making them thicker – which is sometimes the approach taken in engineering.

And there is another fundamental human need that is underpinned by natural diversity – our food.

Eating Diversity

Everything we eat is grown from species of animal and plant that were once wild – or, in a few cases (marine fish, for example), that still are. Centuries or even millennia of selective breeding have generated the many super productive varieties that have enabled food output to expand to keep pace with a growing population and rising living standards. As our population continues to rise, and takes on increasingly affluent dietary patterns, more food will be needed.

The aim of producing more food will need to be achieved at the same time as protecting soils, conserving water, hanging on to most of what remains of forests and other natural habitats and reducing the nutrient enrichment of the environment. It will need to be done while coping with the consequences of climate change and the pressures that will come from pests and diseases gaining more resistance to the chemical weapons we have used against them.

We often assume that the answer to this complex problem is some form of technology, perhaps in the form of new pesticides or genetic engineering. It seems that genetics will indeed be an important key to all this – although the real solutions may turn out to be less in genetic engineering and more in genetic diversity.

Ever since we humans first took the step from hunter-gathering to cultivation, farmers have bred animals and plants so as to develop and hone the best possible characteristics for the conditions they faced. Disease resistance, the ability to withstand drought, varieties that can tolerate cold and types which can grow in salty soils were all developed by farmers in different places at different times and across a range of crop plants.

After thousands of years of selective breeding there is, as a result, among our main crops a wide diversity of varieties. On top of this are the wild relatives of the species we have chosen to domesticate.

They still live in the wild, where they have continued to evolve the solutions to survival challenges. So long as this cultivated and wild diversity is maintained, we have a unique resource to fall back on in times of change. Without it, we are more vulnerable.

Bananas are a case in point. This is the most popular fruit in the UK. We eat about 6 billion of them every year. But these elongated fruits of a tropical plant, packaged up in their bright yellow skins, are all basically identical.

Thousands of years ago, in a steamy tropical jungle, a farmer stumbled across a type of banana that produced a rare kind of fruit. Unlike most other banana trees this particular one had large, sweet fruit. It was a natural mutation and whoever found that banana tree realized it was something special and decided to reproduce it. But there was a problem: the mutation also caused the fruits to be sterile – they produced no seeds. So in order to maintain the special tree, cuttings were planted, and then cuttings were taken from that first cutting, and so on, right up to the present, when we are still taking cuttings.

Because there has been no exchange of genes among the commercially important banana plants for a very long time, the trees which produce the fruits that grace our fruit bowls today are effectively identical – virtual clones. This situation has emerged because the varieties we like to eat are grown from pieces of plant, rather than from seeds that arose from a sexual encounter. This method of reproduction has taken our favorite fruit out of evolution. While the original mutant might have had some resistance to disease when it was first found thousands of years ago, over time its ability to withstand attack can be expected to decline. And that is indeed what is happening. Our bananas have remained static in a genetic sense, while the disease-causing organisms that would attack it have not. They have continued to shuffle, mix and hone their genetic makeup so as to maximize their success, and now the ever more defenseless bananas are under attack.

For a lot of people it's a matter of life and death. In many developing countries bananas and the related plantains are a staple subsistence food, important for the nutritional wellbeing of about half a billion people. This is why the recent effects of disease epidemics caused by two species of fungus have been compared to the Irish potato famine. Banana production in the Amazon has been devastated and there are fears for food security in parts of Africa. One response is to use more potent fungicides more often. And that is indeed what has happened. Bananas are notorious for heavy chemical application with up to forty sprayings per year normal on some plantations. The development of new chemicals could buy time, but the fungi will continue to evolve and might soon gain resistance to our next chemical counter-attack.

Another possibility is to look at the pool of genes held in the still wild relatives of the commercial varieties we have come to rely on. In the case of bananas, that approach has borne some fruit with new strains under development that might be better able to withstand the relentless fungal onslaught. This has been achieved with a technology called genetic transformation, which enables plant breeders to move genes within species, even when no flowering and pollination is taking place. New genetic material will inevitably be needed to maintain food security through the development of new varieties of other crops too. But we can only use what genes actually exist, and that is why in recent years people have gone to some extraordinary lengths to hang on to what is still there.

While heated debate on the extent to which the future of farming should be founded on genetic technologies continues, there is no doubt that genetics will be vital for our food security. Not, however, so much in the form of the genetic engineering and inserting of genes from other species into crop plants that has been so controversial in recent years, but more through the conservation and development of the genetic heritage that is held in the accumulated diversity of crop plants.

Passed to us by Nature and honed for thousands of years by billions of farmers, the value of the vast collection of biological ideas held in crops and their wild relatives might well prove to be the most important resource we have in maintaining healthy human populations long into the future. The manner in which new technologies might be harnessed in this way was recently highlighted with the 2005 announcement that scientists had sequenced the genetic code of rice. As the process of gene sequencing becomes mechanized, so more and more species are being understood in this way.

By understanding the full genetic makeup of this vital plant, scientists will be better placed to develop rice varieties fit for our new world of rising population, changing climate, water scarcity and diminishing natural resources. Through a better understanding of which genes and combinations of genes confer which advantages, we can speed up the selective breeding that farmers have practiced for millennia. But this is, of course, only the case if the genes are still there.

Until recently, the genetic diversity of crop plants was preserved by farmers growing different varieties in their fields. They kept and shared seeds, and in the process maintained varieties and bred new ones. The many different varieties were maintained because they had particular advantages – for example, in resisting pests or being tolerant to drought. But, with the rise of the modern intensive farming that has helped food production to keep pace with population increase has come paradoxically a major threat, that could ultimately make the problem it has sought to solve even worse.

This is because one consequence of growing ever more intensive monocultures of higher-yielding crops has been to see many old varieties falling out of use. According to the UN's Food and Agriculture Organization, about 75 percent of the genetic diversity present in crops in 1900 is lost already. It's gone. And much of what remains is at risk, including that which might have a vital role in assuring our future nutrition.

This point has not passed unnoticed, however, and in recent years there have been renewed efforts to collect, catalogue and store examples of the many varieties of crop plants and some of their wild relatives. Today some 1,400 "gene banks" are in existence across the world. Some focus on particular crops, while others are based on countries and regions. Most are chronically underfunded and some vulnerable to disasters.

In September 2006, Typhoon Xangsane became headline news as it swept through the Philippines, Vietnam and Thailand. The storm killed several hundred people and caused hundreds of millions of dollars' worth of damage. A wall of floodwater and mud also struck the Philippine National Plant Genetic Resources Laboratory. Many of their samples of peanut, sorghum and maize varieties were salvaged, but others were not, and crop varieties that had been developed over centuries were lost for good. This is why gene banks seek to ensure that duplicates of all their specimens are stored at other facilities.

Work is also underway to ensure that a third copy is also maintained – at the safest facility on planet Earth: the Svalbard Global Seed Vault. The Svalbard archipelago is located in the Arctic Ocean to the north of Norway. It is a remote and barren place, populated by polar bears, and seabirds, and, although no center of civilization, it was chosen as the location for a truly unique collection. Built to withstand all kinds of natural and manmade disasters, the vault was established to maintain the diversity that is now widely accepted as essential for our future survival. Reminiscent of a James Bond set, the concrete and steel entrance to the high tech facility sits incongruously in the Arctic wilderness.

Through this doorway, high above sea level, is the way into a huge mountain vault. The Svalbard facility was built to stand the test of time and to be resilient in the face of climate change, war and politics. The cold climate ensures that, even in the event of prolonged power cuts, the precious seeds will remain cold and secure. It is a rare example of apocalypse planning, and how we humans have prepared in advance a response to global environmental change.

It was opened in February 2008, at which point some 268,000 distinct seed samples from 123 countries were deposited. Two years later and the vault contained more than 500,000, making it by far the single largest repository of crop diversity anywhere; 500 seeds of each variety are kept in the vault, so more than a quarter-billion samples are maintained there. The eventual goal is to maintain samples of every variety of every major crop plant from every country on Earth.

While the agricultural scientists continue to scour the world's farms for new varieties, the same goes for their colleagues in the rangelands, dunes, forests and reefs who document more and more "new" species of wild plants and animals. In the year or so it took me to write this book, hundreds more species of plants were described and about 7,000 more insects too. New species of squid, fruit bat, parrot and frog were among the many other newly described life-forms also discovered in that time.

These species, novel to the eyes of scientists, were added to the total of around 1.8 million or so species that had already been given a name. Estimates as to the actual total (that is, including those that have not been named yet) vary widely, but go up to 100 million. One recent calculation suggests it is between 8 and 9 million – but even that is about six times what we have thus far described.

Specimens of the 1.8 million that have been named are stored in various collections around the world. One is at Tel Aviv University in Israel. This is a small country, and the collection of wild animal and plant specimens held there is correspondingly modest, certainly compared to those held by some other institutions, such as the Natural History Museum in London. It nonetheless contains a gargantuan number of individual specimens.

One corridor on the fourth floor of the Natural Sciences Department has thousands upon thousands of glass-topped specimen drawers, each containing between dozens and hundreds of insect specimens, including flies, beetles and butterflies and moths. Racks of glass slides hold specimens of tiny mites. There are jars with sea creatures and the skins of animals and birds. Store rooms hold several private collections of insects that were donated to the university. They have arrived in boxes, chests and cases, including exquisite carved collection cabinets with glass topped drawers that slide in and out to reveal the thousands of butterfly specimens held within. The cabinets reek of naphthalene, the active ingredient in mothballs, little piles of which are left in the drawers to stop pests attacking the priceless contents. The toxic heaps paradoxically look like mounds of virgin snow.

Some of the specimens kept here were collected in the nineteenth century and have been maintained as part of an irreplaceable archive. Those marked with a red label are especially precious. These are the so-called "type specimens" – the first example of a species to be found and described and the one against which future specimens are compared.

David Furth scans the cabinets with an expert eye. He is a taxonomist (a scientist who studies and classifies the variety of life) and is normally based at the Smithsonian Institution in Washington DC. He has come to Tel Aviv to help sort out the collections and to plan how they might be kept up to date. He pulls one of the drawers back so that we can look at the specimens through the glass top. He puts on his spectacles to study the contents: "These are flies that look like bees. You can tell they are not bees because they have only two wings, not four. As bees decline, we might need to know more about these creatures, because like bees they are pollinators." The cabinet contains about seventy specimens.

Furth sees this and other collections as not only a record of the Earth's diversity, but also as snapshots of times and places: "Each of the specimens has a label to say where and when it was collected, and that is incredibly precious information, because it says something about that particular place on that day." By maintaining such collections it will be more possible for us to chart the impacts of climate change and other trends.

In our fast changing world, looking after collections like this is a task that has never been more important, but which is receiving less and less support. There are fewer and fewer people with the qualifications and skills needed to curate natural history collections like this. In Tel Aviv there are only a handful of people, and they are mostly in their sixties and seventies.

In maintaining the benefits provided by the Earth's biological treasure chest, in keeping open our future options as to how we use it, it perhaps wouldn't be a problem that we have only charted a portion of it, if it were secure and open to documentation at our leisure. But it isn't. As is the case with crop varieties, the diversity of wild organisms is disappearing, and fast. Although the present very high rate of extinction cannot be accurately quantified because of the very same uncertainties which limit our ability to say how much diversity there is to start with, we know that the variety of life is declining rapidly as some of the most diverse areas continue to be cleared, ploughed and polluted. On top of all that there are now the consequences of climate change. From the rainforests to the coral reefs, the Earth's life systems are under relentless and growing pressure.

The decline of the cone snails vividly illustrates the point. Many species of these animals are under threat because of collecting, habitat damage and climate change. With each species that goes extinct, its unique potential to aid the relief of human suffering goes with it. But at least we have begun to understand how these animals can help us. Not so with some others.

Gastric brooding frogs developed the unique ability to host their tadpoles in their stomachs, thus cutting down the number taken by predators. This strategy required some clever physiological adaptations to prevent the adult's gastric juices from digesting their own young. Researchers believe the method used by the frogs to do this could inspire new drugs for the treatment of the peptic ulcers that cause misery for tens of millions of people. We can't do that, though, and because of a rather fundamental problem: the gastric brooding frogs are now extinct. They were lost recently, due to the destruction of their habitat. They disappeared before we could work out how they performed the amazing physiological feats that were so central to their particular set of survival solutions.

And in the face of warnings from both science and the real world it appears that our ability to take the long view is rather limited. We humans, or at least those of us who live in the technologically advanced consumer societies of the West, dwell in the here and now, at the ticket machine we arrived at earlier. While the journey that brought us here is increasingly well known, less discussed is the journey forward, to the next station and beyond. Looking at that hypothetical onward journey, there is no doubt that the diversity of life which brought us this far will be vital in making our next steps. Without it, our ticket might take us no further.

We can of course maintain some of that irreplaceable diversity in gene banks, like the facility on Svalbard, and more of it in zoos, at least temporarily. But there are problems with this. Taking combinations of genes out of evolution means that they become static and lose their ability for dynamic adaptation. There are practical problems, too, arising from the sheer diversity of life on Earth. Even if all the zoos were to focus only on the preservation of endangered species, just a small proportion of those in decline could be accommodated. And seeds cannot be kept indefinitely in gene banks – periodically they need to be grown again to produce new seeds; otherwise the genes decay and the seeds die.

There is a greater drawback still. The genetic wisdom expressed in natural diversity is only one aspect of why life on Earth is so important for our continued welfare. On top of the vast store of information and solutions that could power engineering, design, farming, medicine and energy systems long into the future is a huge complexity of relationships and system.

These keep the web of life functioning, and in the process support the world of humans.

For many ecosystems it seems that diversity is an important characteristic in keeping them working and able to withstand shocks and to recover from them. As species are lost, so the ability of natural systems to properly function goes down.

As John Allen found in the design of *Biosphere 2,* it is not only the things that we need to conserve – the genes, the species, the habitats and ecosystems – but crucially also the relationships that life and its support systems have developed.

One such set of relationships, and which are crucial to how the natural world works, are those that have evolved between the flowering plants and the animals that enable their sex lives.

CHAPTER 4
THE POLLINATORS

$1 Trillion: Annual sales dependent on animal pollination

$190 Billion: Annual services provided to farming by animal pollinators

Two-thirds: Major crop plants that rely on animal pollination

IN DECEMBER 1884 *the steamship Tongariro set sail from London en route to Christchurch in New Zealand. A new iron hulled vessel with a displacement of over 4,000 tonnes, she was built for the long distance work required to keep a global empire connected, and possessed the new innovation of refrigeration. Demand for meat and dairy products was growing, and supplies from the far-flung colonies could be more easily shipped back to the home country with such facilities. On the way out to New Zealand, however, she carried in her cold bay an altogether more unusual cargo: bumblebees.*

Farmers in the new colony of New Zealand had quickly realized that the moist temperate climate was ideal for pastures on which to rear sheep and cattle. But there was a problem. The clover with red flowers that they had brought from England, and which so enriched their new pastures, could not produce seeds in the new southern hemisphere territories. Seed had to be imported from Britain so that new pasture could be sown each spring. The reason these economically vital plants failed to reproduce themselves, and why expensive and inconvenient seed shipments had to be made from Europe, was because the clover lacked the natural pollinators that enabled its flowers to set seed.

It was Charles Darwin who realized that the tube-shaped flowers of clovers relied on long-tongued bumblebees for the plant to complete its life cycle. They had evolved together – flower and insect a perfect match for each other. The clover flowers needed the bees to move pollen, while the bees needed the nectar supplied by the flowers for food.

The shipment of bees in the refrigerated compartment of the Tongariro was not the first attempt to transport live English bees to the other side of the world. An "acclimatization society," set up by New Zealand colonists to engineer their environment so as to render their new home more like the old country, had already taken delivery of several batches of dead bumblebees. Whole nests had been packed up into boxes and kept warm for the voyage. But the bees had died. Then shipping just queen bees was attempted. Packed in moss, they got damp, mold set in and they died too. The refrigerated facilities on the Tongariro presented a new opportunity. Short-haired bumblebees were collected by Kentish workmen charged with ditch clearing duties. They were offered a bounty for any live queens they could find. In this way bees were collected and gathered together for the month long journey to the other side of the world.

The bees arrived in Wellington and were shipped on to Lyttelton to arrive at the gardens of the Canterbury Acclimatization Society on January 8, 1885. Of the 282 bees collected in Kent, 48 were still alive and released into the wild in a part of the world that had never known such animals. The already mated queens flew off on the midsummer breeze and helped to found new dynasties of bumblebees. And they did it fast. In1886, bumblebees were reported 100 miles to the south, 86 miles to the west and 55 miles north of the release area. By 1892, bumblebees were so abundant that it was feared they would be a problem for beekeepers because of the competition they would cause to honey bees.

In all, eight foreign bee species were deliberately released in New Zealand, so as to make honey and to pollinate crops and pastures. These insects have been economically vital. All of the country's exports of meat, dairy products, forest materials, fruit and vegetables and wool depend to some extent on insect pollination and bees have thus been a major part of the wealth creation of New Zealand – one of the world's richest countries.

But why is it that some plants require insects to produce the things that we rely on?

Plant Sex

As with higher animals, flowering plants have male and female anatomy. The structures that produce pollen are analogous to the testicles of animals that make sperm. The organs fertilized by the minuscule grains of pollen are comparable to the eggs produced by animal ovaries. When the genetic material contained in a grain of pollen is united with that in the nucleus of the egg, then a viable seed can develop. The act of pollination undertaken by the bumblebees in New Zealand was completed when they moved the pollen from one flower to fertilize an "egg" in another. When they did that, then seeds grew.

Why plants go to all this trouble, and entice insects and others to help them, is for the same reason animals go to a lot of trouble to have sex. And that is because sex enables the individuals participating in it to maximize the survival opportunities of their offspring and thus their own genetic makeup.

Offspring and seeds produced by sex have a mixture of two parents' genes. Without sex, reproduction takes place by cloning – and the result of that is identical individuals possessing only one set of parental genes. Some animals and plants are forced to use the cloning method at least some of the time, for example, when a mate cannot be found, but the advantages of sex are so overwhelming that the vast majority of higher animals and plants invest a great deal of time and energy toward perpetuating themselves via the sexual route.

The most advanced and specialized strategies to ensure that pollination takes place have been developed by the world's flowering plants. Before the emergence of these advanced organisms, which now dominate life on the land, terrestrial environments were still green, but with more primitive plants, including liverworts, mosses, ferns and forests of trees that were the ancestors of modern conifers.

In the period before animal pollinated plants evolved their specialized structures and strategies, pollination by insects and other animals was probably still taking place, but by accident rather than design, as creatures that fed on the nutritious reproductive organs of plants moved pollen around. This might have been the first step toward the evolution of the flowering plants we live amidst today. All four of the main groups of pollinating insects, the beetles, flies, bees and wasps and butterflies and moths, had evolved before advanced flowering plants, so it seems that something happened to trigger the new relationships that are so prevalent today.

The key switch that appears to have taken place is in how the insects went from being predators of the plants' reproductive organs to being deliberately attracted by the plants. Initially this was probably through the plants making extra protein rich pollen, which insects would come to eat, and in the process move to different flowers, thereby completing the sexual ambition of the plant. Evidently the disadvantages that came with making extra pollen were outweighed by the advantages that came from more successful sexual encounters.

This seems to have first happened about 140 million years ago (about three-quarters of the way to Cambridge on the imaginary journey we took in the last chapter). Around then, representatives of the family of plants that includes the modern magnolias, nutmeg, cinnamon and avocado appeared. The group that includes tea, chocolate, cotton, pumpkins and melons among its modern representatives was also present. It was the beginning of an explosion in diversity that would ultimately lead to there being today about 400,000 different species of flowering plant. While some (such as the grasses) rely on the wind to move pollen between different individuals, and a much smaller number on water, the vast majority of flowering plants – about 90 percent of modern species – depend on animals to transfer pollen between flowers.

The reason pollination by animals makes sense is quite straightforward. Broadcasting pollen on the breeze is a haphazard and inefficient method. Finding ways to take pollen directly from one flower to another is a more reliable way of making sure the job gets done, and requires less pollen. It does, however, require other investments, including making the structures to attract the animal pollinators, such as nectar producing organs which make the sugary substances that so many animals love. Plants also need to invest energy making the flowers and scents which advertise that the nectar is on offer. If the plant is successful in enticing an animal to come and feed on the sweet liquid, or on excess pollen, then there is a good chance it will become dusted and then carry the tiny genetic capsules to a different flower, enabling it to grow seeds.

To begin with, it was probably a rather broad set of relationships that developed between plants and animals in this way, with "generalist" insects visiting quite a wide spectrum of plants. But as time went by, more and more sophisticated relationships emerged, with some animals being linked to only one or a small group of plants. Different kinds of insects, and indeed some birds and bats, have adapted for feeding specifically on high energy nectar – and little or nothing else.

Studies from tropical forests show that – as might be expected – plants which are highly dispersed tend to have their own specialized pollinators, while more common mass blooming trees are often pollinated by a variety of generalist insects.

One group that is often associated with highly specialized pollination is the orchids. These plants, with their wonderfully intricate flowers, have developed ways of attracting insects that go beyond the normal reward of nectar, with as many as a third of the 30,000 or so species in this group achieving pollination by deception. They lure pollinators to their flowers with a promise of food, but don't actually provide any.

Bright colors on the flowers, sometimes with sweet scents, are enough to get the insects there and, although disappointment follows the insect's visit, they will have picked up the pollen. The plants evidently bank on the assumption that they will soon get fooled again, and deliver the pollen they first picked up to a different flower that is using the same trick.

Another group of orchids take the deception a step further. They use a strategy known as sexual deception. Male insects are attracted to the flower through the release of a sexual pheromone used by females to entice their mates to come and find them. The male attempts to mate with the flower and in the process picks up pollen. He then takes the pollen to the next false mate on a different flower. This is a very sophisticated strategy and means that those plants using it tend to have only a single species of insect that does the job of pollination for them.

While we tend to associate pollination with bees, there are something like 100,000 different kinds of animal pollinators out there. The majority are insects and a great many are indeed bees, but birds and mammals are also included.

I have watched a lot of birds over the years and some of the most gorgeous are those that undertake pollination. In a South Australian eucalyptus grove, I saw a flock of musk lorikeets (a type of small parrot) feeding with a group of New Holland honeyeaters. Both species feed partially on nectar and pollen and are adapted for that job. The lorikeets have brush-like tongues while the honeyeaters have longer decurved bills that can reach into flowers. I watched for more than an hour as the beautiful red, green and blue lorikeets tumbled through the dense flowers, acrobatically clambering around with their feet, feeding as they went. The black, white and yellow honeyeaters hopped around with them, completing a colorful and remarkable spectacle, all laid on by a huge tree that wanted to spread its pollen between different flowers.

In the Colombian Andes, in an area of subtropical cloud forest, I watched seven species of hummingbird visit a single flowering bush. The iridescent birds, some marked with bright purple and shining emerald feathers, dashed and darted between the flowers, their hovering flight and long bills perfectly adapted for a lifestyle dependent on gathering nectar. In return for the high energy fuel that powers their furious pace of wing beats, they spread pollen for the plant.

Spiderhunters and sunbirds are among other bird groups important for pollination.

When it comes to mammals it is the bats that are by far the most important pollinators, especially various kinds of fruit bats. Plants adapted for bat pollination tend to produce white petals and a strong scent so that the animals they wish to attract can find them in the dark. Plants that rely on birds tend to have red petals to advertise their nectar and rarely develop a scent because it seems so few bird species rely on smell to find plant food.

The many, many thousands of pollination relationships that have been established are a key factor in shaping the character of natural systems. Without pollinators, most ecosystems would not function as they do now, and that would lead to a reduction in diversity and a diminishing of the services that natural systems provide for us. Indeed, some ecologists now believe that a breakdown in pollination relationships has been an underestimated cause of past extinctions, as reduced genetic diversity within a plant species can reduce its ability to adapt to changing conditions or to develop resistance to diseases.

While these ecosystem-sustaining aspects of pollination are important for the functioning of the human economy because of the services provided by ecosystems, such as flood protection and carbon storage, of most practical and immediate importance for us is the contribution that pollinating animals make to agriculture.

When the Wild Bees are Absent

As became apparent with the arrival of European style farming in New Zealand, pollination is vital for food production, both for the plant crops we eat and the productivity of the pastures upon which meat and dairy producing livestock is grazed. But it is not just a problem of the past. Across the world, there are farming communities who are aware of the practical value of pollinators – because of their absence. One example is the almond growing region of the Central Valley of California.

This is one of the most intensively farmed regions in the United States; in an area of more than a quarter-million hectares, some four-fifths of the world's almond crop is produced. The conditions are ideal. It is mild, there are cool wet winters, the soils are right and there is plenty of sunshine when the almonds are growing. So good is it that the land is cranked to the maximum to boost both yields and profits.

Robin Dean is an adviser on strategies to increase bee populations, and with his wife runs a successful enterprise called the Red Beehive Company, which helps clients ensure effective pollination. He explained to me that the exclusive farming of almonds in the Central Valley had brought the land close to crisis: "The soil is being hammered, water is under stress, there are issues with soil salinization from irrigation, and then all the issues with crop protection chemicals. Almonds are harvested mechanically with a tree shaker that causes the nuts to drop to the floor. The nuts are then pushed into rows, are left to dry and then they are vacuumed up. With this system you don't want anything growing on the ground under the trees. The entire area is completely barren except for almond trees and it is like that for miles." One consequence of this, says Dean, is that "there are no natural pollinators remaining in the landscape." And this poses something of a challenge.

In early spring vast rolling expanses of white and pale pink blossoms emerge, and on warm days it is vital that the flowers are pollinated so as to put in motion the growth of the almonds. What happens in this short window determines the size of the final crop, and the difference between successful and patchy pollination can be measured in many millions of dollars. This is why the growers pay a fortune for beehives to be brought in for the six weeks or so in which pollination must occur. And they need a lot of them.

Chinese workers pollinate fruit trees by hand.

Over a million hives are required to pollinate the almond trees in California's Central Valley. They are trucked in from all over the USA to join this annual pollen fest. To get ready for the great event beekeepers move their colonies into staging areas close to the almond orchards, and then as the blossom breaks, the bees are moved amongst the trees in an operation mindful of a military maneuver. There are generally between about 40,000 and 80,000 bees in each healthy hive, and each one pollinates around 300 flowers per day.

As might be expected, the rapid increase in the area planted with almonds has led to an increase in demand for pollination services, and that in turn has led to an increase in the price of hiring bees to pollinate your trees. Today, almond growers pay about $200 per colony to rent bees during the pollination period. Thirty years before, they paid around $10-12. Some of the businesses that meet this demand are running very substantial organizations. "One operator runs about 80,000 colonies for hire to pollinate crops. They are not getting a lot of honey out of this but making money from the pollination," says Dean.

When the almond flowers have withered and with the seeds set and ready to grow, many of the bee colonies depart for the north, to Montana, North Dakota and to Oregon and Washington. Spring comes a little later there and the bees arrive in time to help pollinate the cherry, apple and pear orchards.

In other parts of the world, more extreme measures have been required to ensure that trees bear fruit. A case in point can be seen in Maoxian County in Sichuan, China. Fruit farmers there have had to resort to more direct action in filling the gap left by the loss of natural pollinators.

This part of China lost most of its pollinators back in the 1980s, and now people have to do the job themselves, by hand. In spring, as the trees burst into flower, thousands of farmers climb in the branches of their apple and pear trees. Using brushes made of chicken feathers and cigarette filters, which they touch against the flowers, they hand pollinate the blossoms, transferring the sticky grains of pollen from one flower to another.

The main problem in this part of the world was the excessive use of pesticides. As Robin Dean explained, "In China it is chemicals that have obliterated the natural pollinators. In the foothills of the Himalayas it is cool and honey bees are not an option for fruit pollination. Bumblebees would be a natural pollinator but their populations have been wiped out. They might bounce back but in the interim about 40,000 people have to do the pollination by hand." So, while pesticides were deployed to keep yields high in the expanding orchards, the opposite effect was in fact achieved.

As was the case with clover in New Zealand, there have been other occasions when farmers have encountered problems with pollination when moving crop plants from one geographical area to another.

The oil palm is native to West Africa and in the early 1960s there was an attempt to set up new plantations in Malaysia. Climatic conditions were ideal and the palms thrived, but they didn't produce much fruit. It soon became apparent why this was – the pollen from the male flowers was failing to reach the female ones. As in China, the first step was to employ expensive and laborious pollination by hand.

While this far from ideal solution was put in place, researchers discovered that in Cameroon, where these plants are native, pollination is achieved by a tiny beetle. Following a period of careful screening this creature was introduced into Malaysian oil palm plantations in 1981. The cost of pollination fell to nearly nothing, while within five years fruit production increased from 13 to 23 million tonnes. In just a few decades palm oil output had rocketed, not least because it is the most productive use of land in the regions where it grows. Without the pollinators, its economic viability is, however, massively compromised.

These little pollinators have enormous practical significance, not least because palm oil is in quite a lot of the products we use: from margarine to shampoo and from biscuits to ice cream. Given how widespread this crop is spread across different products, the chances are that you will be a beneficiary of that little beetle at some point today.

These and other examples where wild pollinators have for one reason or another been absent provide us with powerful reminders as to how fundamentally we rely for our wellbeing on animals mediating the act of plant sex. This has generally been regarded as a free service, and until recently it has largely been that: indeed something we can reliably take for granted. However, there has lately been cause to question this view, as fundamental changes to ecosystems have led to growing concern over the continued reliability of pollination.

The extent of our dependence on pollination services is underlined by the remarkable fact that some two-thirds of the different species of food crop plants are pollinated by animals. These different crops produce about one third of the total calories we eat, not to mention most of the vitamins, minerals and antioxidants that we need to remain healthy.

The United Nations Food and Agriculture Organization (FAO) is one of the global specialist agencies that undertakes research to support world food security. It estimates that in 146 countries about 100 species of crop plant provide 90 percent of the food supply. Of these, 71 are pollinated mostly by wild bees, with others pollinated by different insects, including flies, moths and beetles. Pretty much all the blueberries, grapefruits, avocados, cherries, apples, pears, plums, squashes, cucumbers, strawberries, raspberries, blackberries, macadamia nuts and dozens of others depend on the foraging activities of bees. No bees, then no fruit – or at least a lot less.

It is not only the quantity of fruit that is reliant on pollination; there are also issues of quality. Watermelons that have more frequent visits by pollinators tend to have darker fruit with a richer flavor. There is evidence that pollen carried by bees over a long distance may have a measurable impact on the quality of coffee.

And there are significant quality issues linked with pollination in apples. If they have no other option apples will self-pollinate, but if forced to this last sexual resort the fruit quality is compromised. Robin Dean did a study on this: "We covered some apple flowers with bags so bees couldn't get at them. On each tree we chose identical sets of blossom but put a bag over one set and left the other open, marked so we knew where it was. We then looked at the difference in the fruit development. In the set where the pollinators had access we had a higher mineral content and the fruit was about 10 percent firmer. This had implications for storage. When apples go into storage you have steady deterioration in quality, with firmness going down over time. With this difference you had up to six to eight weeks longer than in a cold store. So you don't need chemicals to keep the fruit, which saves money."

And even though apples will set fruit without pollinators, they produce less of it. Robin Dean described to me another experiment that he had undertaken so as to better understand what happens to fruit yield when pollinators are absent: "We had 125 blossoms in two separate sets, one with access for pollinators and one not. In the group where the pollinators had access, we got 60 apples; in the other one, we got 30. So we still got apples, but only half as many and they were different."

Even among those food plants where it is not the seeds or fruit that we eat, but the leaves, stems and tubers, pollination is still vital for seed production. And at the same time as pollination helps to produce seeds for next year's crops – including beetroot, parsnip, carrot, lettuce, onions, leeks, swedes, turnips, sugar beet and all kinds of broccoli – the process also helps maintain the genetic diversity in plants, and that in turn helps preserve their adaptability in the face of environmental changes, including shifts in climate.

Linda Collette is a crop biodiversity specialist at FAO. She points out how the world's pollinators are still largely underappreciated: "Because insects are so inconspicuous, or perhaps because the system worked fine without much intervention in the past, the level of general public awareness, or even specialized awareness among farmers and agronomists, remains quite low. The fact is that ecosystem services provided by pollinators are essential for food production and contribute to the sustainable livelihoods of many farmers worldwide."

So what is the economic value in financial terms of the work done by pollinating animals? There are a few estimates out there. Alexandra-Maria Klein, an agroecologist based at Germany's University of Göttingen, says that the crops relying on animals for pollination account for about $1 trillion of the world's $3 trillion annual sales of agricultural produce. The reason why only about a third of total agricultural sales are down to animal pollinated plants, when most species of major crop plants are pollinated by animals, is because of the huge importance of a few species of wind pollinated grasses in our food system – such as wheat, maize, barley and rice.

Another way to approach the question is to calculate the cost of replacing the services provided by pollinators. In answer to this, an international process hosted by the United Nations Environment

Program called The Economics of Ecosystems and Biodiversity (TEEB) concluded in 2010 that the value is about $190 billion.

While these kinds of global figures help put into context the overall contribution of pollinators to the human economy and food security, very often of more practical interest are studies into the value of pollinators in more specific circumstances. One case study produced for the TEEB process looked at the value of bees' pollinating activities in Switzerland, a country dotted with neat orchards and market gardens. It concluded that colonies kept by Swiss beekeepers ensured annual agricultural production worth about $213 million.

Although honey and beeswax are the most obvious outputs from beekeeping, the Swiss research estimated these as only a quarter of the economic value of pollination. It also highlighted a major gap, in there being no policy then in place to make sure pollination continues. While governments would not consider neglecting or spending on power networks and transport infrastructure, the "green infrastructure" was taken for granted.

Another piece of research, undertaken by a partnership between global crop product giant Syngenta and the World Resources Institute, looked at the value of pollination provided by bees to blueberry farmers in Michigan, USA. It assessed the work of bees to be worth about $124 million annually. Blueberries are often characterized as a super food; they are high in vitamin C, rich in fiber and contain substances which protect the heart and have anti-cancer effects. These health benefits would not be possible to deliver without the help of bees.

These and other studies which have looked into the economic importance of pollination have recently come to be of far more than academic interest. This is because there is now highly convincing evidence to show that pollinators are in worldwide decline.

Monoculture Blues

During the early 1990s I took a car journey from central Poland to Berlin. As we traveled through Poland the car windscreen washers and wipers were in frequent use to clear dead insects that had collided with the car. Winged creatures of all shapes and sizes became splattered on the glass. But a remarkable thing occurred on the border. Once across the River Oder and into Germany, we hardly needed the wipers at all. The reason was we had passed between very distinct agricultural landscapes.

Whereas in Poland there were small fields, woods and wetlands scattered through the landscape, limited use of farm chemicals and a relatively low level of mechanization, on the German side the landscape seemed virtually sterile. In common with many other intensively farmed parts of the world, this region of Germany had witnessed a drastic reduction of wildlife, including many insect pollinators. This was, of course, deliberate, as industrialized farming has sought to exclude many life-forms other than the crop plants themselves. And when it is largely wind pollinated cereals that are under production, the loss of pollinators will hardly be noticed.

Very few farming systems can work without animal pollinators, however, because it is rare that it is only cereals that are produced. And there are certainly no natural or semi-natural terrestrial ecosystems that can do without pollinators and still retain their normal functions and full spread of diversity. As we saw in the last chapter, diversity itself often has value, and to retain that it is important to maintain the many relationships that exist between them, including pollination.

There are very few measures of overall pollinator decline, as we don't have a lot of data on changing insect populations. There are exceptions, however, notably in relation to honey bees in the developed world. In Europe and North America the number of honey bee colonies is known to have gone down, while declines in wild bee colonies have also been recorded.

Many European butterflies are also under serious threat. According to Butterfly Conservation Europe, about one third of European butterfly species are in decline with about 10 percent at risk of extinction. The main reason is the loss of flower-rich grasslands and wetlands, through agricultural intensification. And the impacts of climate change will make matters worse.

Among mammalian and bird pollinators, at least 45 species of bats, 36 species of non-flying, pollen-spreading mammals, 26 species of hummingbirds and 7 species of sunbirds are included in the ever lengthening list of those deemed as either at threat of extinction or past that point and already gone.

In many parts of Europe and North America the decline of bumblebees has been especially marked. The causes for this seem to vary from place to place, but a combination of habitat loss and disease appear to be at the root of most declines. For example, the loss of open grassy habitats such as rough pasture and hay meadows, replaced with arable fields or intensive silage production, has taken a toll. So has the shift in how many gardens are now set out with decking and lawns of pure short grass, leaving fewer flowers for the bees. In parts of the USA, a parasite appears to have made a major impact.

In the UK, where there is reasonably good data it is known that over the last seventy years two species of bumblebee have become extinct, while six of the remaining twenty-four are listed as endangered.

Beyond all the specific circumstances relating to the fortunes of different pollinators, if there is one thing that unites the trends behind their decline, it is the shift toward large-scale monoculture agriculture. This is, of course, the same style of farming that is causing the soil damage described in Chapter 1, the progressive nutrient enrichment of the environment set out in Chapter 2 and the mass extinction of animals and plants covered briefly in Chapter 3. Here we are in Chapter 4, and another major ecosystem function is under pressure from the same quarter.

As was the case on my journey through Poland, when farms are small family-run affairs, pollinators can often survive in patches of natural habitats: the woods, rough pastures, hedges and the little patches of wetland, for example. But as massive industrial farms have replaced such fragments of relatively natural areas with vast fields, so the pollinators have been eradicated, or at least put out of reach of the crops. This is why the European honey bee has become so important.

Friendly Fire

It is not only the almond orchards of California's Central Valley where honey bees are such vital economic actors. Right across the globe these insects are essential workers in the fields, and for some crops they are as indispensable as the machinery, and indeed the farmers themselves. Adapted for living in the cramped conditions found in a tree hollow, honey bees take well to compact wooden hives and to being trucked around the countryside to visit different crops. Which is why, of around 20,000 different bee species in the world, we have come to rely very heavily on just a handful.

These species now dominate the global pollination business and their economic value is no secret. Indeed, they have become heavily, even brutally, industrialized. Robin Dean described the process: "Bees are flown to the U.S. from Australia. You can buy a box of 1.5 to 2 kilograms of bees. There are about 10,000 to the kilo, so you have a good idea of what you are getting before you pay. They are loaded into a box with a syrup feeder to keep them going and are shipped over by air freight to California. They then get loaded into a hive with their queen and they are ready to start work."

When it comes to honey bees, the vulnerabilities that come with reliance on a single species (effectively a monoculture of pollinators) have recently been demonstrated in the widespread phenomenon of so-called colony collapse disorder. This process sees honey bee colonies lose adult insects over a period of weeks until the colony ceases to exist.

One piece of work by Rabobank, a big lender to farm businesses, concluded in 2011 that the numbers of U.S. bee colonies failing to survive each winter had gone up from an historical average of about 10 percent to over 30 percent. Across much of Europe the proportion of hive losses rose to about 20 percent. A similar pattern is emerging in Latin America and Asia. But why is this happening?

Many possible causes have been examined. They range from the effect of mobile phone proliferation (unlikely) to disease and the effect of various pesticides (very likely). Some 175 different agricultural chemicals have been found in beeswax, so these substances, many of which are designed to kill insects, certainly come into close contact with bees. No wonder that beekeeping associations have called for urgent reviews of the safety of chemicals that have come into use at the same time as colony collapse has become frequent – including various neonicotinoid pesticides. Of all the unintended consequences that arise from how we treat Nature, the loss of pollinators caused by pesticides is one of the more ironic. Chemicals that were developed to protect agriculture are undermining its viability. A case of "friendly fire" if ever there was one.

In addition to the effects of chemicals, the collapse of colonies may also be linked to the effects of monoculture farming, including a lack of food for the bees to eat. Robin Dean believes that "food is a big issue ... quite simply they are starving. And like many other species, when you go into starvation mode, stress rises and all sorts of other problems emerge, such as more disease."

Dean explained to me that honey bees eat more than nectar and highlighted the importance of pollen grains in their food: "Nectar is made of fructose and sucrose and fairly standard, but pollen is something else. Different species of plant produce pollen with different compounds and trace elements, and these might be influencing behavior. Different pollen has different protein content, some high and others low. You need a mixture and this is a big issue with monocultures. There might be a lot of pollen in a field of oil seed rape, but it is only one kind. It has only about 11 percent protein, whereas some others are about 37 percent. In a monoculture it is all or nothing, feast or famine." He believes that this in turn might affect the functioning of colonies: "When the crop is flowering, the workers effectively instruct the queen to lay more eggs to produce more workers. She does this, and then the flowering is over and there is nothing to eat because of the time lag between her laying and worker hatching, which is 21 days. The oil seed rape will be halfway through flowering before she increases the egg production. This causes a crisis and then disease can take over." Whatever the causes of colony collapse disorder, it is increasingly seen as a strategic issue for food security. The researchers behind the Rabobank study noted how the output of animal pollinated crops has continued to increase (going up fourfold since the early 1960s) and how this had been achieved with fewer and fewer bees. But it struck a note of caution: "Farmers have managed to produce with relatively fewer bee colonies up to this point, and there is no evidence of agricultural yields being affected. The question is how much further this situation can be stretched."

This is a good question, and one to which there is no definite answer. Given the fundamental importance of food security for our overall wellbeing, it suggests a more cautious approach is warranted.

The speed with which bees and other pollinators have been lost from many parts of the world underlines how quickly a crisis might emerge, as do those places where pollinators have already been lost – such as the fruit orchards of Sichuan, China.

Bee-Roads

Moving more honey bee hives around the countryside is one way to ensure crop pollination continues. Another is to conserve and encourage wild pollinators. There are now many projects underway across the world to restore populations. One of them was set up in the UK by the Co-operative Food supermarket chain, in partnership with the conservation group Buglife. It involves the identification of corridors in farmed landscapes where so called "bee-roads" can be established. As their name suggests, these will be areas of good-quality habitat along which bees can move. Key to the idea is the re-establishment of areas that provide plenty for the bees to eat. That means flowers, including species such as knapweed, scabious and red clover, in turn attracting and supporting honey bees, hoverflies, butterflies and moths. It's too early to say whether it will work, but it should do if sufficient areas of good quality flower meadow can be created. The flagship bee-road project aims to establish two long rows of wild flower rich habitat stretching from east to west and from north to south across the entire length and breadth of Yorkshire – England's largest county.

The Co-op is not alone: among several consumer brands that have also invested effort into bee conservation is major retailer Marks and Spencer. Individuals, too, are stepping up, with more and more amateurs taking on the hobby of keeping hives, including in urban areas. Garden flowers provide an important source of nectar, while the hives assist with the pollination of vegetables, fruit bushes and trees.

There are also governmental efforts underway to restore populations of wild pollinators. One is along the shores of Lake Michigan, where the blueberry harvest is so important to the economics of local farming. In recognition of the millions of dollars of value provided by wild pollinators, in 2007, the U.S. Department of Agriculture's Farm Service Agency launched a new scheme whereby growers could decide to devote some of their land to creating habitats for pollinating insects, in return for some modest financial reward. In the target area, which includes twenty-two Michigan counties, farmers can apply for money to pay for the creation of flower rich grasslands and bare patches of soil that are favored by some ground nesting pollinators. Bees, butterflies and moths are among the groups that the scheme is intended to especially help.

As is so often the case in restoring the services provided by Nature, there is no single solution that will work everywhere. In some agricultural landscapes it will make more economic sense to adopt strategies other than taking land out of production for the restoration of flower rich areas. Some, like Robin Dean, suggest that a more holistic approach is also needed, so as to reintegrate wildlife back into farmed landscapes in a more fundamental way. "We need more work on the benefits of wildlife, and not only bees," he says. "For example, higher diversity pasture produces better milk, and that in turn would be good for pollinators. We need to remove the blinkers and stop seeing things in isolation. We need a broader approach and to educate farmers to see the variety of Nature as a good thing, rather than as automatically bad."

As part of the process of seeing beyond industrial monoculture, and in developing a more profound view of how farming might work with Nature, Dean says that it would be positive to increase awareness about how pollination works in the real world. To make his point he describes the experience of one of his clients, a pear grower in

Portugal. "He was importing honey bees to improve his crop but found there was no improvement in the yield. What he didn't take into account was the fact that down the road there was a eucalyptus plantation. The honey bees were flying right past the pears, because they are low in nectar, and down the road to the eucalyptus trees, to tank up to the absolute gills on a really high sugar crop."

Dean realized what was going on and came up with an alternative. "We suggested that he clear the surface of the soil to make bare patches, which the mining bees love. As mining bees moved in, the crop yield jumped by about 12 percent. It is a question of understanding what is going on and engineering the landscape to suit the particular pollinators you need. People generally understand that bees pollinate crops, but they don't tend to know which bees pollinate which crop."

All these measures make sense – whether it is recreating suitable habitat, or developing more integrated farming in which wildlife is a part of the land use, or moving away from monocultures of domesticated honey bees and toward the re-establishment of robust populations of wild pollinators. They will help protect ecosystems and the relationships that hold them together, and also reduce the risks that come with increased reliance on very few pollinator species.

Dean estimates that in the UK, much of which is intensively farmed, there are enough honey bee hives available to do about 10 percent of the pollination work. "Even if you take a hugely optimistic view, bearing in mind all the hives that are moved around, they still do under 30 percent of the pollination work. The other 70 percent is being done by creatures other than honey bees. Among other things it is the solitary bees, the mining bees and bumblebees."

Aside from the protection and restoration of the habitats needed to sustain populations of these creatures, in some cases more drastic action is needed to restore their fortunes. One of the bumblebee species that has become extinct in the UK was the species released in New Zealand, and whose story there opened this chapter. Once widespread across southern England, the short-haired bumblebee was last seen on the wild flat Dungeness peninsula on the south coast of England in 1988. It was officially declared extinct in the UK in 2000, its demise brought about by habitat loss and agricultural chemicals.

But in this case, the process is set to be reversed. Following a period of habitat restoration on Dungeness, short-haired bumblebees were imported – from Sweden and in a different color variation – and released in early summer 2012.

Albert Einstein famously said that "if the bee disappeared off the surface of the globe, man would have only four years to live." While even the most apocalyptically pessimistic ecologist would struggle to sustain this line of thinking, this genius saw more clearly than many the fundamental importance of pollination in sustaining the human economy.

Fortunately there is no need to test Einstein's prediction. We clearly possess the means to keep the world's pollinator populations strong and robust, if we want to. All we have to do is invest in the many practical and often simple steps that will take us in that direction. Everyone who has even a small garden can help with this. One thing that might encourage us all to do something is to remember how over a period of years the value of the pollination work being done by animals is worth literally trillions of dollars, and not only in relation to food production, but also in sustaining the many natural ecosystems that provide us with so many other services, including (and as we shall see in a later chapter) the supply of fresh water.

While bees and pollination have been in the headlines during the last couple of years, there has been rather less attention to some other aspects of how our welfare rests on pillars of ecology, including in the control of pests and diseases.

Dogs feed with vultures by the side of the River Ganges.

Chapter 5
Ground Control

$34 Billion: Costs associated with the loss of vultures in India

$310: Annual value per hectare of pest control by birds in a coffee plantation

$1,500: Annual value per hectare of pest control by birds in a timber-producing forest

ONE BRIGHT CLEAR MORNING in April 1993 I was aboard a British Airways 747 that was preparing to land at New Delhi Airport in India. As the aircraft started its final descent from about 1,000 meters, I noticed vultures. Their broad wings, with flight feathers spread out like a span of long fingers, had taken them to the height of the plane. From this vantage point their incredibly sharp vision would enable them to scan the land for the dead animals they feasted on.

Riding thermals just a few hundred meters from where I sat, the huge soaring birds flashed by the windows. It crossed my mind that one might get sucked into an engine, as from time to time they do. Aside from immediate safety worries, I was fascinated to see such a number of large birds of prey flying over a built-up sprawl of houses, roads and industry. I was impressed by their abundance – little did I know that these birds were in big trouble. About a year earlier vulture numbers in India had begun to go down – and fast.

Three different kinds of vultures are native to India and all of them had recently declined at a truly unprecedented rate. There were about 40 million vultures in India in 1993, but by 2007 the population of the long-billed vulture had dropped by nearly 97 percent. For the oriental white-backed vulture it was even worse, with a 99.9 percent drop in numbers during that same period. In other words, it had gone virtually extinct.

The decimation of the vultures was inadvertently brought about by a new anti-inflammatory drug called diclofenac. Initially developed to treat people, it was also used to help livestock. The immediate, often dramatic beneficial effects it caused in sick animals made the drug very popular with country vets. But if animals are treated with diclofenac a few days before their death, traces of the drug remain in their bodies. Vultures live from the dead bodies of other animals. So when they fed on cattle, water buffalo and other carcasses, the drug went into the vultures' bodies. It poisoned them, and then they died.

Matters were made worse by the fact that dozens or even hundreds of vultures feed on a single carcass, so when a dead animal has traces of diclofenac it can affect a lot of birds in one go. As a result, all three Indian vulture species are now listed as critically endangered. The oriental white-backed vulture has gone from being what was thought to be the most numerous large birds of prey in the world to one of the most imperiled. Its population at the time of writing is believed to be no more than about 10,000 birds.

While at first sight the loss of vultures might be of interest mainly to bird conservation groups, it was soon found that a whole lot of other people were affected. The question of whom and by how much was investigated by a research team led by Anil Markandya. His team used interviews, field surveys and various official information sources to find out how much the near disappearance of the vultures had cost the Indian economy.

One set of costs was borne by the very poor people who used dead cattle remains as a source of raw materials. When cows fell dead, they were skinned by people supplying the leather industry. With the skin removed it was easy work for the vultures to quickly dispose of the flesh, leaving the bare bones drying out in the hot sun. The bones were then collected and sold as feedstock for the fertilizer industry. Not nice work, but for very poor people at least a source of income.

Putrefying bodies of large animals left out in the sun are a public health hazard, and with no vultures to clean them up they were burned or buried. In this way the disappearance of the vultures deprived many poor people of at least part of their livelihood. It also created new costs for communities, who had to stump up for carcass disposal. Even if any carcasses were left out, the bones that remained were generally of poorer quality because other scavengers, such as feral dogs, don't clean them as thoroughly as the vultures.

The loss of the vultures also created concerns for faith communities, notably the Parsi. The practitioners of this faith believe fire, water, air and earth are pure elements that should be protected from contamination. This is why for thousands of years they have laid out their dead in the open air at remote hill locations called "Towers of Silence." Vultures frequented these places and quickly reduced corpses to bones. With their loss, the Parsis have had to resort to other means for disposing of their dead, including solar concentrators which produce high temperatures and reduce a body to bones in about three days. Such devices cost about $4,000 each.

The costs felt by poor working people through the loss of raw materials and by the Parsi in making alternative funeral arrangements are minor, however, compared to the consequences for public health. Rotting carcasses provide a rich breeding ground for bacteria, including those that cause disease in humans, such as anthrax. The decline of vultures has also created more space for other scavengers – such as rats and dogs.

Researchers looked at dog population estimates and found evidence that the decline in vultures had been accompanied by an increase in dog numbers. Chris Bowden helps to coordinate a coalition of groups called SAVE (Saving Asia's Vultures from Extinction). He gave me an impression of how much cleaning work the vultures were doing: "The 40 million or so vultures that lived in India during the early 1990s were between them eating about 12 million tonnes of meat a year, mostly from dead cattle but also some water buffalo."

It was estimated that this quantity of food, made available after the vultures had gone, would sustain around 4 to 7 million more dogs. And that appears to be exactly what happened. Between 1992 and 2003 the dog population is estimated to have gone up by about 7 million. And, while the estimated increase in dogs does not prove a causal link between declining vultures and rising dogs, the fact that official sources also indicated a relatively stable dog population between 1982 and 1987 (when there were still a lot of vultures) suggests that the two trends are indeed related.

And these dogs are feral. They receive no veterinary care and carry many diseases, notably brucellosis and distemper, which can be spread in livestock and to humans, and rabies. India has the highest rate of rabies infection anywhere in the world, and dog bites are the main way people contract it. Over 95 percent of fatal cases start that way.

The researchers found evidence for an increase in rabies in rural areas during the period when the vultures declined with infection rates heavily biased toward the poorest people. Rabies had been partially controlled in India with vaccines, but in the years after the vultures declined, when the dogs increased, there was a rise in people seeking post-exposure treatment – that is, people who had been bitten.

Markandya and his team of researchers attempted to put some numbers on this. They estimated that in the period 1992 to 2006, nearly 40 million additional dog bites occurred due to the increase in dog numbers. These, they estimated, led to between 47,395 to 48,886 additional deaths from rabies, compared with if the vultures had still been there. This is a remarkable conclusion: nearly 50,000 people are estimated to have died because of the disappearance of the vultures. Remarkable too is the estimated impact on India's economy. Between 1993 and 2006 the loss of the vultures was estimated to have cost the country about $34 billion.

These numbers have not gone unquestioned. The work of Markandya and his team relied heavily on public data, some of which was incomplete. Their findings also rested on a good deal of statistical analysis, and so the numbers I have cited here, and which came from research, are open to challenge, and the authors agree that more work is needed. On the other hand, the researchers built in assumptions that tended to make their estimates more conservative, so the true costs could actually be higher than those quoted. They certainly dwarf the economic gains from improvements to cattle farming that came with diclofenac.

Even if the true cost was an order of magnitude lower than their figure of $34 billion – let's say $3 billion – it seems to me that it would still be economically sensible to do what the researchers suggested. That is, to ban the diclofenac that killed the vultures and to begin a large scale captive breeding program, so that the birds' former numbers might be restored. New drugs which do not harm vultures are now available to treat farm animals, while the cost of a proper vulture captive breeding program over twenty years would cost less than $100 million.

There has been some success in cutting diclofenac use. Chris Bowden says: "A ban was imposed and it was made an imprisonable offense in India to make and sell vet formulations of diclofenac. But companies are still making the human version of the drug, and selling it in vet-sized bottles. So its use is down, but it's still being used at a level that will prevent the recovery of the vultures."

It is perhaps appropriate that some of the garbage wagons that take away household waste are called vultures. They get a bit smelly in the summer, but, as has been the case with their namesake birds, if they don't turn up people soon notice. One big difference, of course, is that when the birds arrived for work, they provided their services for free.

Diluted Diseases

It appears that wildlife can also play more subtle roles in disease control. In this respect, and as was assumed in the design of *Biosphere 2,* diversity itself has value. One way this works is through a process called dilution.

Humans are afflicted by many different pathogens, including some which were once associated more with animals than people, such as Ebola, avian influenza, bubonic plague, anthrax, Lyme disease and West Nile virus. These and some other serious diseases reproduce by infecting us humans, as well as a number of other animals.

Some of the diseases that infect both people and animals live in mammals; others, in birds. As well as being transmitted directly, many infections between animals and humans occur via so-called vectors – such as mosquitoes. Simply put, when there are more potential targets for a vector like a mosquito, and when some of the hosts they might bite do not make particularly good hosts for the microbes that cause the disease it is carrying, then the spread of the disease among humans can be hampered by diluting its chances of reaching a human host.

This appears to be the case in relation to Lyme disease. Research into the spread of this tick-borne condition suggests that it is less likely to reach people where there is a higher diversity of small mammals. The mice, deer, foxes and all the rest of the woodland creatures that inhabit the same ecosystems as the ticks seem to "absorb" more of the potential infections, meaning that there are fewer encounters with humans. Some of the animals have natural immunity, and so don't get as sick as we would, but because the ticks have bitten them they don't need to bite a person to feed.

A similar effect seems to work in helping cut human infections of West Nile virus. This disease was first identified in the West Nile District of Uganda in 1937, first appeared in North America in 1999 and became a significant cause of human illness in the U.S. in 2002 and 2003. It is spread by mosquitoes (including those that feed on birds), and causes big problems for humans, with inflammation of the brain being the most serious. The way infections spread during the U.S. outbreak of 2002-2003 appeared not to be random, however, and John Swaddle and Stavros Calos carried out detailed research to see if there was a relationship between the level of wild bird diversity in a particular area and the level of human infections.

Their research relied on comparing differences in the bird diversity between neighboring U.S. counties, and where one county reported West Nile virus, and where the other one didn't. Efforts were made to take account of climatic factors, variations in the kinds of mosquitoes found, the level of affluence in particular areas and how urbanized the different pairs of counties were. In cutting through all these variables, the research did indeed uncover a relationship between the number of bird species present and the level of human infection.

Where bird variety was higher, the rate of human infection was found to be lower. With more types of birds, the mosquitoes that transmitted the virus had more chances to feed, and were thus less reliant on humans to complete their life cycle.

The researchers concluded that the level of bird diversity in different areas could explain about 50 percent of the variation in human infections. This is a dramatic finding, especially considering how much money is spent by public health authorities in controlling outbreaks of disease like this, and how much misery they cause. In the ten years after it arrived in the U.S., West Nile virus killed more than 1,100 people. In 2002 alone, the healthcare costs associated with dealing with West Nile virus was estimated at $200 million.

The message seems quite straightforward: more diverse wildlife provides us with a buffer that can sometimes curb the spread of disease, and the value of that service is considerable.

Tits and Apples

Even dedicated birdwatchers rarely think of birds as essential economic actors. In many instances, however, this is exactly what they are. In addition to pollination services, waste disposal and disease control, some of those pretty birds are among the many creatures that undertake economically vital jobs as pest controllers.

For some years I have tried to encourage great tits to breed in a nest box in one of the gnarled old apple trees in our small Cambridge garden. While our offer of purpose built accommodation was declined, a pair did in the end set up home in a hole right next to the nest box, but which had been excavated by a great spotted woodpecker. When the chicks hatched, the adult birds began the breathless task of trying to find enough food for their fast growing offspring.

We'd sit in the garden and watch the spectacle of the two parents flying back and forth, one or the other of them arriving at the nest hole every minute or so with food. A closer look with binoculars revealed how for the most part they were bringing back caterpillars. When the family finally left the nest hole in the apple tree and flew around the garden, we discovered there were seven babies in there.

On the basis of our very basic observations, some simple arithmetic generated surprisingly big numbers. At that time of year (May) and at the latitude of Cambridge, the daylight lasts about seventeen hours. Admittedly, I was not watching these birds all the time, but whenever I did they seemed to be bringing food at an ever more furious rate as the chicks grew. Assuming that the adult birds worked for only half the time, this is still eight hours – or 480 minutes per day. At a rate of about one feed per minute, that means nearly 500 insect grubs were taken to the nest each day. The chicks were fed in the nest for about twenty days and so it seems reasonable to estimate that some 10,000 caterpillars and other food items were caught and eaten. More were consumed as the parents continued to feed the chicks outside the nest, and added to that is, of course, the consumption of the adults themselves.

It struck me that these handsome and acrobatic little yellow, black and green birds, with their sharp beaks and sharper eyes, were undertaking a rather important job on the fruit trees in both our gardens and neighboring gardens. They seemed to hunt in those trees where they expected to find most food, and that included the fruit trees, which at that time of year were covered in growing insect larvae. If left unchecked, these would, of course, by the end of summer, spoil a lot of the fruit.

I didn't know how much, and therefore had no idea what the actual beneficial impact of the birds (if indeed any) might be. Fortunately, however, a couple of Dutch scientists did some work to find out. Christel Mols and Marcel Visser measured the level of damage to apples in an orchard where great tits were nesting, compared to orchards where they weren't. To increase the number of nesting great tits in some of the study areas, they put up boxes. Unlike the birds in my garden, they readily moved in.

By contrast to my anecdotal garden observations these researchers gathered a lot of proper data. Caterpillar densities were measured by a forensic examination of the trees. In the autumn they checked the apples in the different experimental areas to measure how many were spoiled by insect larvae. Damage caused by caterpillars in spring was easily identified through the corky scar tissue left behind. The results demonstrated a striking effect. In the areas where the nest boxes were used, the number of apples damaged by caterpillars was reduced by half. The areas with the most great tits also had the best harvest of high quality apples.

The authors concluded that reducing caterpillar damage in orchards by offering nest boxes to great tits is an extremely low cost measure that can increase the harvest of undamaged apples by more than a tonne per hectare per year. Less pesticide is needed, and that is of benefit to the fruit grower (in cutting costs) and of course beneficial for other wildlife (in not being poisoned).

Fruit and vegetables are not the only beneficiaries of pest control duties carried out by birds. The supply of one of our favorite beverages is also in part down to birds. Work carried out in the Blue Mountain highlands of Jamaica shed some light on how much. Matthew Johnson of Humboldt State University, Arcata, California, along with two of his colleagues, examined the effect of birds in cutting down insect damage in a shaded coffee plantation.

Coffee was once routinely grown under tall forest trees. Recently it has more often been grown in full sun so as to boost productivity. While additional sunshine can boost short-term yields, shade can, among other things, provide cover and nest sites for birds. The birds need to eat, and what they eat includes the insects that would otherwise damage the coffee crop. In this case the birds were found to have a major impact on the population of an insect called the coffee berry borer – the most damaging coffee insect pest in the world. When the effect of the birds was assessed on the coffee yield (by experimentally excluding them from certain areas with nets), they were found to be providing about $310 worth of value per hectare.

High in the hills of Kenya and Tanzania, Unilever owns extensive tea estates. Among the remarkable characteristics of these places is the fact that they don't use any pesticides. One possible reason why they have been able to get away with this is because of the areas of natural forest and other habitats that have been protected and nurtured since the plantations were set up in the 1920s. The forests also keep the atmosphere moister and provide wood for drying the tea leaves.

Richard Fairburn works for Unilever. He has spent many years in East Africa and knows the tea business there very well. He says the trees have been retained on the estates for good practical reasons, to conserve soils and to help stop fertilizers getting into the rivers. On both these counts the protection of the forests has been a very successful strategy. The trees also help ensure that rivers continue to flow in drought periods and act as corridors that allow wildlife to move freely around the estates, including many kinds of birds. And this in turn might be why there have been so few problems with pest infestations. "When I went to our Kenyan estate in 2001," Fairburn told me, "there was an old planter's wife who thought there were 100 species of birds there. I didn't know if that was right

137

and so decided to get the National Museums of Kenya to do a proper survey. They found 174 and said that if they had more time the true figure might be more like 220."

Given the progressive development of resistance to pesticides by a range of harmful pests, these and other findings are of far more than passing interest. Birds and other predators can evolve in step with their insect prey. Where our chemicals can't easily keep up, the birds can and do. This is a point that is not, of course, only relevant to tea.

Mols and Visser, reflecting on their findings in the Dutch orchard, remarked how they felt the role of birds in pest control had generally been overlooked. They didn't speculate as to why this might be, but in technologically obsessed societies might it be that we more easily relate to a person with an insect spray than we can a small bird going about its daily business? Have we become blinded to the obvious? Or is it that there is more money to be made from selling pesticides than nest boxes? Or perhaps we have approached economics wrongly, in for example being able to put a financial value on the chemicals, but not that of the birds? I suspect it is a combination of all three.

Whatever the reasons, the more we look, the more we discover how a host of services provided by different animals has a huge economic value in pest and disease control. And it's not only in relation to food production that it has been possible to put credible economic estimates on the value of the work.

Frogs, Owls and Lacewings

Wood is vital, not least for making paper and the timber we use in construction. But, like food, wood production can be massively impacted by pests. One source of widespread damage comes from the western spruce budworm. These caterpillars of a common moth infest the buds of various coniferous trees across a large area of North America. When conditions are right they can visibly defoliate large areas of forest, as their numbers grow to plague-like proportions.

Fortunately, a number of songbirds like to eat budworm, especially a large-billed bird called the evening grosbeak. Their impressive mandibular toolkit is adapted for splitting seeds, but in the summer they seek out budworms to feed to their young. And, like the great tits in my garden, they are rather good at it.

Two scientists, John Takekawa and Edward Garton, tried to find out what the impact of the birds actually was and to translate that into some financial numbers. They went into the forest and collected information on what the birds were eating and when. On one particular eleven-hectare study plot, the grosbeak flocks were estimated to have eaten about 9 million budworms in just one month, and to have done this during the period when the pests are at their most vulnerable.

When the work of other birds was taken into account the total budworms consumed on that same area was nearly 13 million. The birds reduced the need for spraying and, according to the researchers, were protecting about a quarter of the economic value of the forests. The dollar value of the birds' work in controlling pests was estimated at about $1,500 per hectare per year – and that was the going rate in 1984. Not surprisingly, this and other work suggests that it would be rational to manage forests with more birds and fewer pesticides.

In early 2012, while travelling in rural Bangladesh, I met some farmers who had reached the same conclusion. In the vast floodplain of the Brahmaputra and Ganges deltas there are tens of millions of farmers, mostly working tiny plots of land and between them growing a wide range of crops, including potatoes, lentils, garlic, onions, aubergines, cotton, tobacco, jute, wheat, various fruits and, above all, rice. They are members of the vast global army of smallholders who produce more than half of the world's food output. It is often suggested that, in order to make these farmers more productive, more chemicals and fertilizers is the way forward.

North of Jessore, in the area around Jheneidah, I spoke with farmers who had been using different means to control pests and protect crops. Over the past few years they had taken to sticking a tree branch in the soil of their rice paddies and vegetable fields. A couple of meters tall, the branches attract insect eating birds by providing them with somewhere to sit while they look for food, including harmful insects among the crops. Forktails, shrikes, bee-eaters and bulbuls were among the insectivorous birds that I saw sitting on the perches in the fields, and it seems that their presence was having the desired effect.

In the warmth of the afternoon I sat in a shady grove of trees with a group of farmers. An interpreter asked in Bangla on my behalf if any of them had seen a benefit from attracting birds to their little fields in this way. One young man stood up and said he had seen a cut in insect pests and as a result needed to use fewer chemicals. I asked if any other farmers in the group agreed with the man who first spoke. There was an excited hubbub of agreement among most of the twenty-odd farmers.

Farmers are notorious for copying what works from each other. The fact that nearly all the fields around where I visited had dead branches planted in their fields said a great deal. It was encouraging to see so many birds in such an intensively farmed region, and indeed measures to actively attract them. But there were no vultures – for the same reason that there are few left in neighboring India. There were plenty of dogs, however.

It is not only our feathered friends who are important agents for biological pest control. Many species of amphibians (such as frogs and toads), mammals (such as bats) and insects (including ladybirds, lacewings and hoverflies) also play a vital part in keeping pests in check. Not only are the services provided by all these animals cheaper than chemicals, but they are also part of the food webs and other relationships which sustain other vital natural services.

Our scientific knowledge improves all the time, but these basic facts of ecology have been known for a long time and inspired many attempts by people to enhance the role of biological pest control. The problem is sometimes one of awareness, however. Robin Dean, the pollination expert who we met in the last chapter, told me about pear sucker, "a pest that has become resistant to virtually all chemicals that can be thrown at it. They are an aphid kind of thing and are a big problem in the early season. They penetrate the young shoots and leaves on pear trees and cause them to die off. There is a species of flower beetle that eats them, though. These creatures overwinter in nettles, so if you have nettles near your pear trees you have more of them in the spring and then more biological control of the pear suckers, and then less of a problem."

Of course, you have to have the knowledge to apply it. According to Dean, part of the challenge is communication. "There is a whole bunch of academics who know this and a whole load of farmers who don't, because they are just not talking to each other. If we are to move more toward integrating this knowledge with technology, then a more sophisticated approach is needed. Instead of using a particular chemical at a particular time of year, it is better to look at the pest level and to decide which is the best strategy to deal with it."

As was found with the great tits in the Dutch orchards, it seems that it's not simply a question of chemicals or how to harness Nature, but really one of how best to use both together, in a process that ideally minimizes one (chemicals) while maximizing the other (Nature).

The economic sense of this is becoming more and more obvious. Dean says that in some parts of the U.S., the cost of crop protection with chemicals now exceeds the value of the crop: "All sorts of things have contributed to this but the result is to be throwing more money at the problem than it warrants in simple financial terms. It would be interesting to sit and take a good hard look at this from the point of view of how a more joined-up and integrated approach might work."

Why we might wish to do this is underlined by the fact of hundreds of insect species that damage crops becoming immune to different pesticides. They have evolved resistance to our chemical concoctions and will continue to do so as we develop new ones. Despite an estimated increase in pesticide use of more than 800 percent between 1961 and 1999, the pests have in many cases developed biological countermeasures. Birds, frogs and other living pest controllers evolve with their prey, however, thereby always keeping up with the creatures they eat. Favoring Nature's pest controllers is thus in many cases evidently the better choice, not only for reasons of ecology, but also for economy.

What this might mean in practice varies by place and pest. Boxes for owls and tits (which respectively eat mice and caterpillars), ponds for frogs and toads, dense cover left to encourage nesting thrushes (which can be voracious snail predators) and flowers to attract various beneficial insects can all make a positive difference. If our small garden is anything to go by, it can be a dramatic one. As has been demonstrated by an increasing body of research, these beneficial effects can be enhanced and scaled up to create effects with huge financial value.

In addition to encouraging natural pest controllers, such as tits and grosbeaks, biological methods of crop protection can be boosted by other means. There are many instances of how farmers and gardeners have deployed predators, parasites and pathogens in their age-old war against pests, breeding and releasing the organisms that might rise to the task at hand.

Small wasps have been successfully used to control whitefly. The wasp lays its eggs on the pest and its larvae kill the host (pest) as they develop. Infestations of red spider mites can be overcome by releasing a different faster growing mite. A recent innovation in slug control involves the use of microscopic nematode worms. These parasitic creatures seek out slugs, reproduce inside them and in the process kill them.

A combination of introduced parasites and careful crop management has helped control the alfalfa weevil. During the 1950s, this very damaging pest was accidentally introduced into the United States, where it required farmers to use large quantities of pesticides. Cutting down on synthetic chemicals not only saved money but also spared the natural predators of the weevils from the ecological collateral damage that accompanies pesticide use. More of them survived to help with further control. But, alongside the positive cases of where enhanced biological controls have worked, there are many disasters.

Mistaken Identity

Foxes were introduced into Australia to control the rabbits that had been released by European colonists. The rabbits had got out of control and, by eating up the sparse pastures that were the basis of meat and wool production, were causing economic damage to the sheep industry. But the foxes evidently didn't realize what they were meant to be doing, and instead of controlling the rabbits, they began to decimate the native mammals and birds, which were in some cases also under pressure from the rabbits, that were eating their food.

The consequences of cane toads being introduced to Australia would prove even more disastrous. These animals, native to South and Central America, were transported for release in newly established sugar cane plantations, where it was hoped they would control infestations of cane beetles. They did indeed do that, and ate a lot of other insects too. But so successful were they in their new niche that their numbers grew fast.

They began to take over the habitat of native Australian amphibians and also brought diseases that caused mayhem among the local species, which had little ability to resist novel pathogens. Even worse than this is the damage caused by the toxic chemicals they produce as part of their natural defenses. When under threat, the toads exude a deadly chemical cocktail, and this has killed vast numbers of snakes, lizards, crocodiles and other predatory animals that have made a meal of one.

A further unfortunate consequence pertains to the cane toads themselves. With no natural controls on their numbers, conservationists have had to resort to running them over, bludgeoning them to death with sticks and jumping on them.

The moral of this and similar stories is clear: don't tinker, not unless all the possible consequences have been looked at, checked, looked at again and all possible better alternatives eliminated. The role an animal plays in one place may not be the role it will play somewhere else, where a different set of relationships have evolved – in the case of Australia, over tens of millions of years of isolation.

There are also examples of how aggressive "pest" control can cause disastrous outcomes. Perhaps one of the most dramatic cases arose from Mao's "war on pests," launched as part of his "Great Leap Forward." Mao issued a directive to do away with all pests and among his top targets for special attention were sparrows, rats, mosquitoes and flies. The campaign demonized Nature as man's adversary, a force to be resisted, subjugated, and finally overcome. As far as the sparrows were concerned, the Chinese were ordered to bang drums, pots, pans, gongs and whatever noise-making implements they could lay their hands on to keep the little birds flying until they dropped from exhaustion. Some impressive results were claimed. In Shanghai alone, the death of 1,367,440 sparrows was recorded.

But the campaign backfired. Released from their feathered nemesis, locusts and other pests were left free to plunder the fields. They devoured crops and people starved to death. This was but one factor in Mao's ill-judged attempt at modernization and industrialization that led to the death of up to 60 million people.

Predators, Prey – and the Climate

The role that predators and prey play in their dynamic interaction between the diner and the dinner extends to an even wider planetary level than that of food production. A trip to the Scottish Highlands sheds some light as to how.

This is one of my favorite places to relax, bird-watch, walk and fish. For me and millions of other visitors, it is tempting to see the impressive mountain landscape as a natural and wild environment. But it isn't. It is what is sometimes referred to as semi-natural; that is, comprised of mainly native species, but not in the form that would have been the case before it was modified by the activities of people.

The hills were once covered with pine, oak and birch forests. There were stands of juniper on the hills and wild cherry trees in the river valleys. These trees still occur but not as they once did, in extensive wild forests. The Highland forest covered approximately 50 percent of Scotland 6,000 years ago. Today it extends over about 2 percent of the country.

The forests were initially cleared by fire and axes to make way for grazing stock and to open land for crops. Today much of the land is used for extensive sheep production. These animals nibble and trim the vegetation, suppressing the forest's return. But even in those areas where the sheep are absent, the forests have for the most part not come back, and the soil has changed, too. The drastic change in land cover from mature forest to open grasslands and heather clad moorlands not only altered the wildlife; it also reduced the soil's carbon organic matter (and released carbon dioxide into the atmosphere).

In Scotland, man certainly played a big part in the decline of the primeval woodlands, but there is another factor in the forest dynamics – and that is the role of lynxes, brown bears and wolves. These animals were wiped out one after the other: first was probably the lynx, gone by about AD 400, then brown bears by around AD 1000 at the latest, and finally the wolf in the early 1700s. They were deliberately exterminated, and with them went the main natural check on the deer population. Today the very high number of deer is the main reason why the forests do not regenerate, and that is of interest to us for many reasons, not least from the point of view of our efforts to reduce carbon dioxide levels in the atmosphere.

A lot of money is spent in controlling deer with fences and guns. This is mainly to protect commercial forestry plantations and to conserve the small fragments of native woodlands that remain. While matters of climate stability rarely enter into discussions about deer control, a significant carbon impact is linked to the loss of the carnivores that once predated the grazers and browsers.

While the polar bear has become a symbol of the damage caused by climate change, perhaps the brown bear might be adopted as the mascot of at least one little bit of the solution?

Beyond Poisons, Sprays and Toxins

With our modern ecological awareness it seems easy to see the obvious failings in Mao's campaign to wipe out pests, and his brutal version of human-centered development. But is our mind-set today entirely different? Rather than finding the best ways to work with natural systems, our continuing assault with chemicals and concrete mostly seeks to overcome, control and subjugate Nature. Given what we know now about the value of Nature and natural systems in dealing with waste, helping to manage disease and control pests, we need to ask why we find it so difficult to protect assets that are so evidently valuable.

Very often the difficulty lies in the differences between who benefits and who loses from particular choices over different time scales, and this in turn is often determined by how powerful and influential the different actors are. In the case of India's vultures, it is poor people who have suffered the biggest direct impacts, while the main beneficiaries of business-as-usual are those who make and sell diclofenac. Meanwhile, the action to support vulture populations has come mainly from a British charity, the RSPB, which is spending about £400,000 per year to start a captive breeding program. Novartis, the company that originally developed and manufactured diclofenac, has expressed no interest in supporting such work.

Waste cleanup, pest control and protection of public health are all services associated with municipal bodies or private companies who we generally pay either through taxes or bills. If Nature stopped doing what it naturally does, then the bills and taxes would need to be a whole lot bigger. And these are not the only public services that we pay for in these ways, and which Nature helps to supply.

Nature is also the world's largest water utility.

Espeletia shrubs in the high mountains of Columbia.

CHAPTER 6
LIQUID ASSETS

505,000 Cubic Kilometers: The amount of rain, snow and drizzle that falls each year

$7 Billion: Savings made by New York City through investing in Nature for its water

0.03 Percent: The proportion of the world's water that is fresh rather than salty

THE CITY OF BOGOTÁ *is spread along a high valley in the eastern range of the Colombian Andes. Home to around 7.5 million people, it continues to grow rapidly, including a series of sprawling slums. In one called Bolívar City, 2 million people live in small houses, packed tight and layered on top of one other. From a distance it looks like a chaotic construction made from children's building bricks. They creep up the sides of the hills — and go higher each year. There is more organized development too, with large areas of new apartments going up for the better off, while off to the north of the city new industrial developments are under construction.*

By 2030, Bogotá is expected to be home to 10 million people, and this will create more demand for all the services that cities must provide: waste collection and treatment, public health services, power, food, schools, transport – and of course a safe and secure water supply.

Some 2,600 meters above sea level, the air is thin and visitors like me who come from low-lying places notice the difference, catching their breath on the stairs. The air here gives a practical reminder of just how shallow our planet's atmosphere really is, a demonstration of the fragility of the envelope of life sustaining gases that cling to our planet.

Higher up still, where it is hard to walk far without becoming breathless, in the forests and open country of the very high mountains, are some of the keys to securing at least one of this city's present and future needs. For above Bogotá are expanses of open country unique to the high mountains of South America. It is known as páramo, a word derived from a Spanish term *páramo*, meaning "desolate territory." Perhaps for the first Spanish explorers it might have seemed a rather bleak world compared to the lush forests and the productive farms lower down. The páramo is very far from desolate, though. It is an amazing living system, and its importance in sustaining the millions of citizens of Bogotá can be hardly overstated, for among other things this is the source of their water.

Head out of the city toward the southeast, past the slums, through mountain villages, up into the pastures, beyond the little homesteads with washing drying in the mountain breezes, onward through the carefully tended plots of potatoes, and the character of the land begins to change. Above 3,000 meters there are fewer big trees and the vegetation tends to hug the land. Further up still, at around 3,200 meters, the feel of the mountains is quite different. This is the Sumapaz páramo. It is the largest area of this kind of natural habitat in the world, and it is breathtakingly beautiful.

The Sun shines brightly, but there is a chill in the air. The altitude gives a feeling like that of tundra. The quality is different to the harshness of the Arctic, though, for despite the raw and exposed aspect of this place it is very close to the equator and there is a year-long growing season. All around is a rich diversity of shrubby plants. Some resemble heathers, others miniature trees. During September many of them are in flower, including the curious looking *Espeletia* shrubs. These plants resemble little palms, with a ruff of grey fronds spreading out from near the top of their stubby trunks, and have yellow flowers that look like daisies. At this altitude they are everywhere.

The landform is dramatic, and there is a brooding ancient presence shaped over millions of years by the action of glaciers and water. It is almost as if a great living being lurks beneath the cloak of green.

The *Espeletia* shrubs, with their pale stems and yellow flowers, are scattered across the undulating terrain in all directions. They give the impression of a soft fur coat on a great sleeping beast, and in so doing present symmetry on a different scale, for on the leaves of the *Espeletia* themselves there is a fine furry surface. Covered with tiny hairs they feel like velvet. Remarkably, this microscopic adaptation is one of the natural features that helps keep the páramo wet to the point where it can supply millions of people with water. Far in the distance, but pin sharp in the clear mountain air, clouds roll in from the west and the Amazon Basin. Great fluffy white formations bubble up above the plateau of the high eastern range of the Andes, their ragged grey bottoms dragged along beneath. The misty lower wisps of the clouds stumble into the land and as they touch the páramo, some of the airborne water is snatched by the velvet leaves of the *Espeletia*. The microscopic droplets in the clouds collect on the tiny hairs, they grow bigger and the water is funneled toward the center of their palm-like rosettes, and then down the stems into the ground. In this way the clouds moisten this land. And there is rain, too — a lot of it. On the Sumapaz páramo, up to 4 meters of rain can fall in a single year. Another example of how the bottom-up meets the top-down.

Jose Yunis works with the Nature Conservancy in Colombia. He is passionate about the páramo. He says that it not only harvests water from the sky, but also stores it for slow release: "This system is like a giant sponge. The sunshine here is fierce and, with the strong winds we get, things can dry out in no time. That is why the slow release of water that comes from the páramo is so important. You can call it green infrastructure."

And that phrase is quite appropriate. For the cloak of green is undertaking a vital job in soaking up and holding water and thereby helping it enter the ground beneath, from where it slowly flows out, maintaining stream flow and water supply. Take away the vegetation, and flash floods and muddy water can be the result. And that is not all the vegetation is doing. Yunis adds that on this massive and remote area of páramo live spectacled bears — the only member of the bear family living in South America. Among many unique birds are Andean condors. These giant vultures, one of the world's largest flying birds, patrol the high mountains in search of dead animals.

Wherever I tread, the land is indeed like a wet sponge. There are carpets of bright green moss and patches of damp grass. The stems of different shrubs emerge from the moist ground. Many are covered with lichens, as well. In one or two of the more sheltered valleys there are patches of more substantial forests with larger trees. Moss, lichens and ferns grow from the trunks and branches. Clustered around tumbling bubbling streams, they line the water courses like a guard of honor. Their presence is more than symbolic, however; they help to protect the riverbanks from erosion and moderate the flow of water.

Some of the streams run into flat boggy areas, where ponds and lakes have accumulated. One of these is the source of the Chisaca River, which flows down toward Bogotá. En route it tops up reservoirs which supply many of the city's inhabitants with fresh water. But once the river leaves the páramo, it no longer runs clear. Around the edges of the upland wilderness there are potato fields and pastures, and these are encroaching into what was, until recently, wild natural habitat. Sheep and cattle wander among the delicate high level forest, where they nibble the baby trees. Farmers collect fuel wood there, too.

Manuel Rodríguez was Colombia's first environment minister and knows the recent history of this area very well. He looks across a valley and explains to me how up until about 2005 the Sumapaz páramo was protected by default through the presence of a FARC guerrilla stronghold: "This place has changed a lot in the last fifteen years. I came up here back then and it was not under so much pressure because the guerrillas were here. But now we have better security and that means more people are coming up to farm the land."

As we make our way higher still on to the páramo, we are stopped and searched by the army. The place is now under state control, and thus safe and ready for investment. My Colombian friends are evidently pleased to see the men in uniform. They all have vivid memories of just how dangerous this place used to be. Everyone knew someone who had a relative kidnapped or murdered. But at the same time they are worried. The once closed areas are now open for development.

Because of the increased farming activity there is evidence of soil erosion. Potatoes have been planted on sharp-ridged hills. Between the bright green rows of carefully tended plants the soil is exposed. When the tubers are harvested, the soil will be completely exposed. Heavy rains will wash the soil into the rivers. Some of it will clog up reservoirs. Some of it will end up in the Caribbean Sea, about 1,000 kilometers distant.

Many of the streams that run down from the páramo, but which aren't protected because they don't feed the reservoirs, are loaded with different kinds of pollution as they pass buildings and urban areas. Some are contaminated with untreated sewage and others receive effluent from various small industrial operations, including highly toxic discharges from small leather tanneries.

Many of the streams have been captured in concrete culverts. These engineering straitjackets isolate the streams from their environment so that they can more efficiently shift floodwater, toxic waste, eroded soil and human excrement. A number of these streams eventually combine to form the Bogotá River, which is then subject to further pollution as it passes through the metropolis. By the time it leaves the city it is putrid, poisoned, black and one of the most polluted rivers in the world.

It flows toward the edge of the eastern Andes, and the steep slope down to the valley of the mighty Magdalena River. As it begins its descent, the river passes through some of the last remaining sub-tropical cloud forest on that side of the mountains. In the gaps between the clouds, and looking west, it is possible to make out the snowcapped peaks of the high mountains in the Andes' central range. They lie about 150 kilometers away, but the air is so clear that they appear closer, creating a disturbing juxtaposition of stinking water and the pristine Nature. The Bogotá River finally joins with the Magdalena on the floor of the valley that separates the eastern and central ranges of the Andes, where it contributes to making this the fifth largest South American river. The Magdalena flows north and sends about 8,000 cubic meters of water into the Caribbean Sea every second.

The Bogotá River is tiny by comparison with the continental monster it feeds, but unfortunately still brings a proportion of the poisonous waste and some of the 200 million tonnes of precious topsoil that the bigger river takes to the sea each year.

Where the Magdalena meets the Caribbean at Barranquilla, a great plume of brown explodes out into the otherwise clear blue sea. I flew over this area and it looked like a great continental toilet has been flushed into the ocean. And that impression is not far from the truth.

The situation is bad enough under normal conditions, but when the Magdalena is in flood, as it recently had been, its brown load is carried far out to sea. The particles and nutrients in the water are swept along the coast on currents, and some of it arrives at the Rosario Islands. These flat little islets are made of coral and surrounded by gorgeous reefs full of many kinds of fish, mollusks, crustaceans and more. I visited there in 2003 and was amazed at the marine wildlife I saw.

But following severe floods which swept along the Magdalena, masses of soil and nutrients were dumped around the islands, and the salinity of the seawater fell as well. All this put huge pressure on the coral reef systems. Part of all this is down to decisions made by farmers high up in the Andes, and the householders and industries lower down who contributed their own bit of sediment or pollution. It provides a powerful reminder of the interconnected Nature of the cycles of water that sustain life on Earth.

One Third of One Percent

This chain of events and causes and effects is of course far from unique. Across the globe our impact on fresh water only grows. Rising demand for the life liquid comes from pressure to grow more food, more affluent lifestyles and the water using devices that come with them, industrial expansion and population growth. All are adding to a rising burden that will cause serious stresses and strains in the years ahead. While traveling along great rivers like the Magdalena can give the impression that fresh water is superabundant, I fear this is another of those feelings that is not as accurate as we' might like to think.

There are about 1.4 billion cubic kilometers of water on Earth. It is thought that most of it arrived in comets that collided with the young planet about 4 billion years ago. Since then it has been endlessly recycled. From clouds to rain to rivers to ocean to clouds to rain to rivers to oceans, it sometimes gets locked in ice and in more recent times sometimes it passes through plants and animals – the endless cycle which replenishes freshwater sustains all the life on land. But the proportion that is in the liquid freshwater state, the water that is so vital for us, is just a tiny sliver of the total.

If it were possible to gather together all the water on Earth, all the seas, lakes, rivers, ice, clouds and groundwater, and to make it into a spherical aqueous blob, it would sit over an area about the size of Eastern Europe. I sometimes use a pictorial representation of exactly this in lectures. I am always struck by how it seems that the little blob of water would barely moisten the great dry rock (the Earth) that it is juxtaposed with. What I find even more striking is how that blob is divided up.

Nearly all of it is ocean, and therefore salt water – a full 97.5 percent is in this state. Of the remaining 2.5 percent that is fresh water about two-thirds is locked up in the ice caps and glaciers. Nearly all of the remaining third is underground in the rocks, leaving only one third of 1 percent of the fresh water on the planet in rivers, lakes, streams, clouds and rain. This vanishingly thin proportion of the total water, augmented with groundwater, is what keeps the global economy going, and sustains all of the variety of life found on land. It is as fundamental for our continuing existence as the air we breathe.

Hardly less significant is the fact that water simultaneously occurs on our planet in three distinct states: ice, liquid and vapor. All three are vital for the functioning of the Earth system. Ice caps, among other things, reflect back solar radiation and keep the Earth cooler than it would otherwise be. Glaciers on high mountains are the source of summer river flow, including through the massive rivers that supply much of the water needed to keep afloat the economies of the two most populous countries on Earth. The vapor state transports fresh water around the planet in clouds that bring rain, snow and mist. The liquid phase enables plants to grow, animals to live and lakes and rivers to exist.

The ultimate source of the fresh water is solar powered evaporation from the oceans. Radiation from our home star hits the water and causes it to take on the form of vapor. This is not the whole story, however. For clouds to form, the water vapor needs nuclei to condense around. Tiny particles of dust and salt spray play an important part, but not enough to explain the vast amount of cloud that forms on Earth each day. Recent research suggests that the process of cloud generation is aided by tiny organisms, including plant plankton called coccolithophores. These miniscule organisms, drifting along at the surface of the oceans, release a chemical called dimethyl sulfide – it is one of the substances that make the seaside smell the way it does. When it's in the air, dimethyl sulfide combines with oxygen to form tiny sulfate particles. Water vapor condenses around these, and that helps clouds to form.

The clouds transport fresh water around the planet, including from the sea to the land. The clouds also contribute to keeping the planet cooler than it would otherwise be, as their white tops reflect the Sun's rays back into space. As a general rule the warmer the sea, the more rapid the algal growth, thus the more cooling clouds that are produced. In other words, the algae are helping to regulate the climate, by producing a cooling effect when it is warmer.

The small and large also combine in recycling water once it has fallen as rain. Clouds spawned over the Atlantic Ocean bring moisture to the east side of South America. A quick glance at a vegetation map reveals, however, that the rainforest of northern South America is not restricted to the coast. It spreads all the way across this vast continent and extends up into the moist subtropical and temperate forests that clothe (or at least once clothed) the slopes of the Andes in Colombia, Ecuador, Peru and Bolivia. Some of the clouds that arrive here were first formed over the Atlantic Ocean, but much of the rain that falls in the far-flung western margins of the Amazon Basin is derived from recycled rain that first dropped further east.

Some of the rain that falls over the rainforest runs into rivers and streams, but much of it is captured by the plants. And as they photosynthesize, they release it again, from those little pores on the bottom of their leaves, which they open to take in carbon dioxide and to release waste oxygen. This process is called evapotranspiration, and in some places it has a major impact on the water cycle. In this case the forest releases so much water that new rain forming clouds are spawned. These drift across the forest, broadly in a westerly direction, taking water further and further from the ocean where rain clouds first originated.

And, as is the case over the ocean, where tiny sulfate particles help to seed rain, the clouds above the forest are in part seeded by plumes of microscopic pollen produced by billions of flowers and carried aloft on warm air. These tiny nuclei cause water to condense into droplets and these eventually coalesce as rain. And some of the clouds that are generated by the forest travel far beyond the Amazon Basin.

Cloud Forests

Many moist forests are thus sustained by water cycles that the forest in large part creates itself. The implications of this for continued water security are clear. If the forests are removed, then there will be disruptions to how water is distributed across large areas. In some parts of the world rainfall is going down as a result of deforestation.

Forests are not only vital in the water cycle because they evaporate water and generate clouds; some kinds of forest also harvest water from clouds, in the process replenishing streams, springs and groundwater.

For example, on the island of Tenerife, in the Canaries, the high volcanic mountain that rises from the deep ocean is often shrouded in cloud on its northern side. The clouds form most days, and at about the same altitude. Although they are often not dense enough to produce rain, they do nonetheless carry a lot of moisture. The mists swirl around the trees and there is a cool freshness in the air, a relief to the hot dry conditions lower down and on the other side of the mountain. The trees are festooned with ferns and mosses, and they drip with moisture. Like the páramo, these forests capture water from the clouds, soak it up, store it and then release it into the ground, streams and springs. This is a cloud forest, and in many parts of the world such ecosystems are vital for water security.

Cloud forests are scattered across many tropical and subtropical parts of the world. They generally occur at above about 2,000 meters and are located on mountains that receive a steady supply of moist air. In many lowland locations it is possible to know exactly where the cloud forests are located simply by looking at the place where each day the base of the clouds meets the mountains. One defining characteristic is the fact that they harvest more water from the air than they release by evapotranspiration. In other words, they are net producers and, in some cases, a major source of water for cities and agriculture.

Professor Neil Burgess is a conservation biologist with a particular interest in East Africa. He has spent several decades working to better understand the forests in that part of the world, including cloud forests that are important for urban water supply. "The forests catch a lot of water but don't immediately release it all," he explains. "So you don't get a huge surge which causes flooding or erosion. The water is stored in the mosses and other plants on the trees and it is trapped in the leaf litter. It can be quite cool up there, so you might find the leaf litter is about half a meter deep and that stores the water too. It's like a sponge and the forest slowly lets out the water it has soaked up."

Burgess remarks on the importance of cloud forests in many countries with seasonal rainfall: "If you are in tropical lowlands during the dry season and it is not raining any more, but rivers are still flowing, the water will sometimes be coming from cloud forests. Even though it's not raining, the forests are still harvesting and releasing moisture, so for dry season water supply they can be crucial."

The Uluguru Mountains in Tanzania provide a case in point. "This is the water catchment for the Ruvu River," explains Burgess. "It supplies the commercial capital Dar es Salaam, which has 4 million people and is an economic powerhouse for the country. In the dry season much of its water comes from this river, which in turn is nearly all replenished from the mountain cloud forests. These ecosystems have huge economic value. Over the last fifty years or so the flow in that river in the dry season has declined by about half. This is probably not only because the forest has been partly cleared – water is being taken for farming and so on – but forest loss is certainly a part of why the river flows less these days."

Water Is Growth

Many tropical countries with seasonally dry lowlands, but with mountains high enough and some moist air coming from the ocean, are in part reliant on cloud forests for water supply. These include fast developing nations in East Africa, a few in West Africa, Vietnam, Cambodia and Laos, the Andean countries and Mexico, among others. There is also a close correlation between cloud forests and wildlife diversity with many areas of remaining cloud forest recognized as hugely important for the large numbers of unique animals and plants they harbor.

In many parts of the world as much as three-quarters of the original cloud forest has gone, cleared to make way for farming or degraded by fuel wood collection. Much of what is left is now in different kinds of reserves, so, while the rate of loss has gone down in many cases, only remnants are left, and often these are under pressure – for example, from grazing livestock.

So, while countries and municipalities have historically tried to keep pace with rising demand for water with a strong emphasis on pipes, concrete and engineering, lately there has been closer focus on the natural systems, such as the peatlands, pastures, forests, cloud forests and páramos that replenish, store and gradually release water.

There are signs that the penny is beginning to drop – even among the governments of major countries: for example, Brazil.

In late 2009, I went to Brasília and met with several ministers from Brazil's national government to discuss how best to design international partnerships that might help with their ongoing battle to control forest loss, including in the Amazon. Whereas in previous discussions some years before they had tended to emphasize their country's right to clear away the forest to make way for development, a different tone was evident. This time there was more awareness about the economic importance of intact forest. This was not only in relation to the huge global role it was playing in soaking up carbon dioxide, but also the contribution it was making to the country's economic growth through its part in the water cycle.

Two factors were behind this change of heart. One was the importance of the rain produced by the Amazon but which falls in the south and center of the country. This is where much of the country's agriculture is located, including many of the huge sugar cane and soya plantations that have contributed so much to Brazil's rapid recent development. With less rainforest, there might be less rain, and that would mean lower exports and wealth being created. Another factor was Brazil's fundamental reliance on hydroelectric power. About 80 percent of the electricity consumed in Brazil comes from dams on big rivers. Recent severe droughts in the Amazon had cut river flow, and that had impacted on power production. The government of Brazil was beginning to see the forest as worth more alive than dead, not least because of the water it supplied.

Slowing down the rate of forest loss there and in other countries is a complex and difficult business, but at least the realization that the forest has a clear economic function might help sharpen political attention and galvanize the commitment that will be so vital for success. Fortunately, this is not some abstract hope. At the end of

2011, Brazil reported the lowest annual level of forest loss since proper records have been kept – with the area cleared in 2010 down to 6,450 square kilometers. This was part of a plan to cut deforestation by 80 percent by 2020, and evidently it was working.

A big part of why this was acceptable politically was because of the economics being seen differently, and in part this was bolstered with incentives offered by the international community, most notably the government of Norway, who are also helping Guyana (as we saw in Chapter 2) and Indonesia.

The work is not finished; however, far from it, in fact. Large-scale development projects are planned for the Amazon region, including dams and roads. There is also political pressure to relax some of the recent controls to allow farmers to further expand production of commodities such as soya. The forest could also be in part lost because of climate change. 2011 saw a record drought in the Amazon Basin and this kind of event could become more frequent as sea surface temperatures in the tropical Atlantic rise.

But why should people in Europe or North America worry about Brazilian droughts or pressure for more deforestation? Where I live, in England, we perhaps suffer from a false sense of security when it comes to water. We have the odd period of prolonged dry weather, but generally plenty of the essential liquid to go around. However, we still have a fundamental dependency on water in other parts of the world.

All of the food and manufactured goods that countries import rely for their production on quantities of water. Tea is a case in point. We drink at lot in Britain but we don't grow any of it. It comes mostly from highland regions in the tropics and subtropics, and where plenty of rain falls.

PG Tips is one of the UK's biggest tea brands, and it is grown in Kenya. Unilever makes it and Richard Fairburn, whom we met in the last chapter, says that tea production in highland Kenya is vulnerable to the effects of continuing deforestation: "Lake Victoria is about 100 kilometers from the Mau Forest, which is in turn next to our tea estates. Each clear morning the sun beats down on the lake and it evaporates. The moisture rises up and travels over the 300,000-hectare forest and becomes even more saturated and makes clouds that produce rain. The tea estate gets rain nearly every afternoon. Without the forest, the clouds would produce less rain and that would be a major problem for us, and indeed the economy of Kenya. We are already seeing signs of change. Over the last twelve years, half have been what we would describe as drought years." One reason might be the clearance of the forest.

It's not only rain-dependent crops, including tea and coffee, that are vital for maintaining the future viability of Kenya's economy. A walk around many supermarkets in Europe reveals another large-scale water dependent export from there – cut flowers. Carnations and roses are particularly important, bringing in many millions of dollars' worth of foreign exchange. One important source of the flowers that grace many Dutch, British and German tables is Lake Naivasha.

This is one of the Rift Valley lakes and is not only vital for flower exports, but also wildlife, tourism, farming and energy production. It is, however, in big trouble. Much of the water that once flowed to the lake from the surrounding land is used to irrigate flower farms. Even though some of the more responsible growers have taken steps to cut water use, the light delicate flowers have a big water footprint. A single rose stem, which might weigh about 25 grams, will have required between about 7 and 13 kilograms for its production. That means each bunch of roses will have needed between 280 to 520 times its own weight in water to produce.

The amount of water used to produce particular products is

called "embedded water." Measuring this helps to establish how much water from a producing country is ultimately being used by consumers in the countries that receive their products. It is a powerful concept.

In the UK, the visible water we use that comes from our taps, showers and toilet flushes totals around 150 liters per day. On top of this is the water used to grow the food we eat, generate the power we use and make the consumer goods we buy. It is rather more than what we use directly – according to one estimate, about 4,650 liters per day.

Making up this total are some everyday goods that in water terms are rather big-ticket items. A kilogram of coffee typically needs about 20,000 liters of water to produce. The tea that grows in the Kenyan hills needs about 2,200 liters per kilogram; 1 kilo of grain-fed beef, about 16,000 liters. Meat generally has a bigger water footprint than vegetables, not least because water is needed at each stage in the food chain, so the higher you go, the more water is needed.

For a country like Kenya, the export of "virtual" water is vital for the continuing output of produce, including flowers, tea and coffee. In the importing countries, all these things make our lives much nicer. And on top of our personal enjoyment, supermarkets make a profit, flower growers maintain a business and people are employed in the coffee and tea estates. The dividends and taxes paid by companies and people working in these industries help to pay for pensions and healthcare. Propping it all up are supplies of clean, fresh water – reliably provided by Nature. Or, at least until recently.

Fairburn sums up the situation in relation to Kenya like this: "The country doesn't have much in the way of oil or precious metals and the economy is dependent on its ecosystems to thrive economically. In fact, all of its major exports are based on water and Nature:

namely tea, coffee, cut flowers, horticultural crops and tourism. The entire foreign exchange earnings are based on the ecosystems. Without sufficient rainfall and water you don't have much left."

The fact that so much embedded water is transported around the world every day makes pressure on the water cycle a truly global question. And it is not only developing countries that are vulnerable to changing patterns of water use and supply. Even in some of the richest and most advanced nations, the distribution of water continues to determine the kind of economic development that is possible, and indeed to influence global food prices.

Travelers heading down from the Outback toward the coast of South Australia cross an important boundary, first identified in 1885 by the colonial surveyor general of Australia, a man called George Woodroffe Goyder. During that year, he drew a line on a map. A plaque set up by the road at Willochra Creek commemorates the fact that the line runs through the little settlement here. In this area, and over a relatively short distance, there are quite noticeable changes in the countryside. The massive gum trees which north of the line grow only in the creeks that take the deluges of rare rainwater off the land become scattered across the flat valley bottoms some distance from the creek-beds. North of the line there is rough and poor quality sheep grazing, with the land dominated by plants characteristic of the vast arid zone of Australia's interior. Heading past the line and to the south, the land takes on a lush and greener character. There are arable fields with cereals and pastures with dairy cows. Some of the creek-beds have standing water. Further south still and some of them actually have flowing water. There are steel grain silos and large industrial scale farms.

It was through these lands of transition that Goyder drew his wavy line. Although an approximation based on limited data and an incomplete understanding of the local plants, his water boundary has nonetheless stood the test of time. For Goyder said that his line was the northern limit of lands he considered safe for agricultural development. Above the line the climate was volatile to the point where boom-and-bust farming would bankrupt families and cause food shortages. He was right. North of his line, brutal droughts have over the years ruined many farmers unwise enough to ignore it.

No one knows if the line might move because of climate change, although some scientists believe that the drought that afflicted much of southern Australia in the early years of this century could be part of a longer term trend of progressive drying. Whether that is the case remains to be seen, but the vulnerabilities that come with the possible shifting of boundaries like this was seen in the contribution to food price volatility caused by the drought which cut Australian grain production during the last decade.

Easier to anticipate are the impacts on commodities like coffee and tea. These grow at particular locations where temperature and moisture conditions are suitable. Climate change projections for East Africa, for example, suggest that the area available for growing tea and coffee might be much reduced in the years ahead, as the conditions which presently make these lucrative crops such a strong source of employment, development and tax revenues begin to change.

Whether linked to climate change or not, all across the world farming is vulnerable to water scarcity. It is not only through plowing more land or using more fertilizer and spraying more pesticides that we have managed to increase food production so dramatically in recent decades; it is also because we have been able to take so much fresh water from Nature as well.

During the past fifty years, the area of global irrigated cropland has about doubled and today some 70 percent of the water we take from the environment is used for watering crops. This rising demand from farming accounts for a large proportion of the tripling in water extraction that has taken place in the period. In many areas, rising demand is met from groundwater, and in various places there is now cause for concern as to how long food output can continue to expand before water availability becomes a constraint.

The removal of water from the ground in China exceeds the sustainable level by about 25 percent, while in parts of northwest India it is believed to be over 50 percent. Both of these countries have vast populations and not only are they growing in that way, but also economically as well – which means more demand for food, and that in turn means more demand for water.

Nature's Sewage Works

The water that is so vital for farming, industry and our domestic uses is endlessly resupplied, with the help of different natural systems, including the oceans, soils and forests. This is not the whole story, though, because Nature also adds millions of dollars' worth of value by cleaning water, too. One place that made a big impression on me in this respect is a large area of wetland that lies east of Kolkata in India. When I visited there in 2004, I was amazed to learn that the wetlands' mosaic of ponds, lakes, channels and swamps were the only sewage treatment facilities the 12 million or so inhabitants of the city then had.

The overpowering stench of human waste hung heavily on the humid air. In the 37°C heat preceding the imminent monsoon, it was almost suffocating. But the smell was not surprising when you know that the wetlands to the east of Kolkata, or Calcutta as the teeming metropolis used to be known, then received nearly 700,000 tonnes of raw sewage every day.

Vegetable plots were established there and, as sewage arrives from the city along slow flowing channels, the solids separate out, the ponds are dredged and the black solids spread on to the small fields and garden plots that intersperse the wet areas. This material provides the nutrients that help the people who live and work on the wetlands produce many thousands of tonnes of vegetables.

Once solids are dredged out, the liquid fraction is passed to pools filled with fast growing water hyacinths. These attractive but tough plants not only remove nutrients as they grow stems, leaves, flowers and seeds, but absorb some of the heavy metals and other toxins released from small-scale industries, such as tanneries. When the organic pollution is partly diminished by the plants, the water is used to top up fish ponds. All across the wetlands there is fish rearing. About a dozen species are raised, in more than 300 ponds covering about 35 square kilometers. Between them, they produce a staggering 13,000 tonnes of fish a year, much of it consumed in Kolkata. Many of the people who live there also keep flocks of ducks. They eat snails living in the fish ponds, which in turn have eaten algae which have grown with photosynthesis, aided by the rich concentration of nutrients in the water.

Some 50,000 people depend on the wetlands for their living: growing vegetables, trading and making nets, or maintaining the channels. The fish rearing alone supports a workforce of about 8,000 people. In a country with widespread poverty, the wetland is economically important. It supports livelihoods and deals with sewage that would otherwise cost many millions of dollars in pipes, concrete and treatment works. This is not to say that Kolkata is a model of best practice for sewage treatment, or that working in such conditions is the kind of development we might wish to see in the future, but it is a reminder that we do have options for working better in tandem with natural processes.

Another large developing country city with major water treatment challenges is Kampala, the capital of Uganda. In 2003, some ninety percent of the inhabitants of the city had no piped sewage system. One way the city has managed to get away with what would otherwise be a serious sanitation crisis is with the help of the Nakivubo Swamp. This area of wetland runs from the central district of the city and through residential districts before entering Lake Victoria.

As is the case in Kolkata, the wetlands help to maintain the quality of urban water supplies by treating and purifying domestic and industrial wastes and effluents. And also like Kolkata, the wetland helps to support some modest economic activities, such as papyrus harvesting, brick making and some fish farming.

An economic evaluation conducted in 1999 suggested that the wastewater purification and nutrient retention ecosystem services of Nakivubo Swamp were worth up to $1.75 million a year. A different calculation estimated that a sewage treatment plant would cost over $2 million per year to maintain. Not only was the cost of expanding the sewage treatment plant greater than the value of the wetland, there were other economic benefits associated with the wetland through the livelihoods it supported. On the basis of these kinds of economic calculations, plans to drain the swamp were reversed and instead, in 2003, the area was incorporated into the city's greenbelt zone. Since then, unfortunately, the wetlands have suffered degradation due to industrial development and other pressures.

Where Trees Make the Rivers Run Fast

The stresses and strains we face in supplying and treating water suggests to many experts that we urgently need to find ways to better understand and reflect the economic value of the natural systems that are of importance in these ways.

In Tanzania, Neil Burgess has seen modest progress in relation to the cloud forests of the Uluguru Mountains that supply water to Dar es Salaam: "One of the jobs we need to do is find ways to create an economic connection in the city with the forests, so that payments are made to keep the system in place. Laws and regulations are needed but people downstream can already find ways to make payments to the people upstream to keep the forests intact. In this case, the Dar es Salaam Water and Sewerage Corporation and Coca-Cola are cooperating to find ways to make payments to upstream farmers. Payments can, for example, help fund alternatives to farming practices that cause erosion and inefficient wood use, including with more efficient stoves to reduce firewood demand."

It wouldn't necessarily strike most people that the way rural people cook could have such a profound impact on the water supplied to city dwellers, but in this case it does.

In Mexico, an official scheme was introduced in 2003 whereby landowners could apply for payments to protect forests in sensitive areas, should they undertake to forgo particular activities such as agriculture and cattle grazing. Points are awarded to recognize the value of particular areas of forest for replenishing water supplies and flood prevention – the more points a piece of land is awarded, the more money is paid to keep its ecosystems in good shape.

During its first seven years, the scheme has attracted participation from about 3,000 landowners and is estimated to have reduced deforestation by about 1,800 square kilometers, which is equivalent to more than halving Mexico's rate of forest loss from 1.6 to 0.6 percent. The benefits are not only seen in the protection of water supplies, but also in the conservation of the wild animals and plants that live in the cloud forests. The scheme is also responsible for avoiding the release of about 3.2 million tonnes of carbon dioxide.

There are many other examples of how water resources can be protected through partnership with land users. I recently saw what I thought was an exemplary one near the French town of Évian. Évian nestles between Alpine foothills on the southern shore of Lake Geneva. Just behind the town rises spectacular Alpine scenery and from this dramatic landscape comes Evian's famous water. It originates on a plateau some 900 meters above the lake, where a mosaic of pastures, hay meadows, wetlands and fields capture rain and melting snow.

These different habitats cover a complex body of geological deposits laid down during successive glaciations, including when the deep broad valley, which is now Lake Geneva, was filled with ice that rose as high as the plateau itself. When rain and snow falls on the plateau, it takes about twenty years before it emerges from a line of springs in Évian below. En-route it takes on the particular mineral content that gives the water its name. The water has given this small town of 8,000 people global fame and sustains the Evian water company, which in turn supports 900 local jobs.

The fact that the water is provided by fragile Nature has not gone unnoticed, and the communities work together to maintain the purity of the water through sustaining the ecosystems that help produce it. More than 100 patches of wetland are scattered across the plateau covering about 10 percent of the water catchment, but account for up to 30 percent of the water that eventually appears in the springs below. The management of the wetlands, and their diverse wildlife, thus brings a clear economic benefit, and this is one reason why they are listed under the Ramsar Convention on Wetlands of International Importance.

Traditional use of the pastures and meadows requires the minimal use of chemicals and artificial fertilizers, and this too helps to protect the purity of the water that has such huge economic and social value.

Some of the money earned by the Evian water company goes to the farmers, to give them an incentive to work the land in ways that will not jeopardize the water.

Bottled water has been a fast-expanding business in recent years. In the UK, the volume sold increased from 500 million liters in 1990 to more than 2 billion in 2009. Not all of this came from beautiful natural areas like that around Évian, but a lot of it did, including some of the other famous brands. Bottled water has been attacked because of the plastic waste it generates and the energy used in shipping it around. Both are fair criticisms. On the other hand, if done right it can be a source of wealth and jobs, based on the careful protection of natural areas.

Either way, bottled water comes in small quantities compared to that needed to supply a global city. But on this scale, too, there are examples of how land is being managed to more cost-effectively supply water.

Manhattan is one of those places that seems as though it couldn't be more removed from Nature, but even in this vast city the water piped to the skyscrapers, apartments and thousands of restaurants serving iced tap water is provided in part through the management of green infrastructure, in this case river catchments in the Croton, Catskill and Delaware hills.

These areas, covering some 2,000 square miles of mainly wooded country, supply high quality drinking water to nearly 9 million people. But maintaining the land so that water is stored and then released in good condition is a complex affair, not least because some 2,000 different owners control the land where New York City's water first falls as rain. A solution was found through bodies set up in 1993 and 1996 to promote water friendly farming and forestry practices.

These set out to encourage voluntary participation in schemes (backed with financial incentives) to promote practices that would not only help support livelihoods but also maintain good water supplies. With money from the New York state authorities and U.S. Forestry Service, it has been possible to involve some 95 percent of landowners in a forward-looking scheme that has protected the city's water at a fraction of the cost of the alternative. The result is the largest unfiltered public water supply system in the United States.

If artificial filtration was installed, so as to strip nutrients and sediments from the water coming from the hills, it would have cost the city around $6–8 billion, and such equipment would also cost up to $500 million per year to operate. By investing instead in best practice farming and forestry, the cost to the city was about $1 billion. This difference was in turn seen in New Yorkers' water bills, which went up about 9 percent, rather than doubling – as was expected had a new water treatment plant been built.

And there is of course a range of other benefits. Much of the catchment is open to, among other things, walking, cross-country skiing and fishing. There are also major benefits for the conservation of wildlife, with woodlands and farms naturally geared up to maintain native plants and animals.

And Bogotá, the Colombian capital we visited at the start of this chapter, might soon travel in a similar direction, through a program of páramo and forest conservation and restoration initiated by José Yunis and the Nature Conservancy. They have helped pull together a group of organizations to work on a long-term plan to conserve the páramo, and other natural features that help to ensure a continuing supply of clean water. The actors involved in this ambitious initiative include Bogotá Municipal Water Services, Bavaria SAB Miller (Colombia's largest beer producer) and Colombia Natural Heritage.

The idea is to restore the green infrastructure that will help secure Bogotá's water supply long into the future. A central aspect of this program is to protect the high mountain páramo. Another will be the reforestation of the degraded farmlands downstream from there. This is not only about conservation of the environment; it is fundamentally about development. If Bogotá's 2 million slum dwellers are to have jobs and economic growth, then water will be one of the key means of achieving it, underpinning industry, ensuring food security and generating power from hydroelectric dams.

And the reason the water company and a beer manufacturer have become involved in conserving natural habitats is because they saw a clear economic justification. They both realized it would be cheaper to protect the source of the water, rather than to rely on engineering solutions and treatment works. For example, one major challenge both of them face is to secure water free of sediments. As the New York authorities found, removing particles of soil is very expensive once they are in the water and it is cheaper to invest instead in preventing the sediments getting into the water in the first place.

Among the practical actions that mark the first steps in this Bogotá pilot program, is the establishment of a tree nursery. Workers employed by the project have scoured the little fragments of native forest left on the hillsides above Bogotá to collect tree seeds. It has been set up amid the ruins of an old stone chapel and its grounds. In there they have managed to work out how to germinate the seeds of 67 species of local trees that grow in the forests of the temperate zone where the reforestation will take place. I saw the different ways they were doing this, with some seeds germinating in big pots of soil and others in shallow bowls of water and moss.

For each hectare of reforested hillside, some 6,000 tree seedlings are needed. There are thousands of hectares of hillside in need of restoration, so a lot of trees will have to be grown. And, as was the case in New York, vital for success will be partnership with the farmers. There are a lot of them, too, and many have just a few of hectares of land.

Reaching agreement on enough of the catchment to make a difference will require a great many individual deals to be struck. The farmers will need to be paid to plant trees in the most sensitive areas and trained in less damaging farming methods. This will cost money, but compared to the benefits gained downstream in ensuring the continued development of a capital city, it will in the end be a small price.

One big question is how to make an economic linkage so that money can flow up the hill to pay the farmers for ensuring that clean water flows down. One way would be through charging consumers via water bills; another could be through allocating a proportion of local tax revenues. There is already a law that requires municipalities to spend 1 percent of their income in protecting water supplies. Although many presently ignore this requirement, with some political commitment a lot of money could be made available for this and similar schemes.

Whether the money needed to scale up the restoration and conservation work comes via the route of taxes or from consumers' bills remains to be seen, but the basic economic point is clear: it is cheaper and more secure to protect Nature than it is to spend money cleaning up sediment-ridden water, or building new reservoirs when a huge natural sponge is already in place and all set to collect and supply masses of water indefinitely into the future.

And there are, of course, fringe benefits that come with this kind of program. For a start, the conservation of the many unique species that live on the páramo, and the temperate forests that once fringed it (and could again). Soil conservation will help ensure food security and the trees will help remove climate changing gas from the atmosphere as they soak up carbon dioxide, as will the accumulating soils beneath the trees.

New sewage treatment plants will also soon help clean up the dire Bogotá River, and that in turn will help clean up both the Magdalena River and the Caribbean Sea. All of this has a financial value, too, although for the most part we've not yet found ways to properly reflect it in mainstream economics.

While the detail of some of the economic mechanisms might still remain in their experimental phases, the water story can at least link the wonderful wild páramo with the cold bottles of beer which enrich Bogotá's fridges and bars – and long may the connection continue. Cheers.

Fish stocks are sustained by a food web that starts with sunshine.

CHAPTER 7
SUNKEN BILLIONS

$274 Billion: Contribution to global GDP from fishing, fish processing and sales

$50 Billion: Extra value that could be gained from well managed fish stocks

$16 Billion: Official subsidies spent in ways damaging to fish stocks

THE CLEAR OCEAN WATERS that surround Tenerife plunge down over the island's steep submerged volcanic slopes to meet the ocean floor some 2 kilometers below the waves. At the bottom of the sea it is dark and cold, but at the sparkling surface bright sunshine beats down. It is high summer and where water meets land there are beaches packed with people on holiday. My son Nye and I have departed the beach and set out to sea aboard a small sport-fishing boat in pursuit of tuna. About 5 kilometers from shore and the sun is intense, the wind light and swell heavy. Eight rods are set in holders on the stern of the boat as it ploughs forward into the rising and falling sea.

Cory's shearwaters, brown seabirds related to albatrosses, patrol nearby. On stiff straight wings they fly low over the water, using light updraughts from waves to help keep them airborne; as they wheel from side to side, their wing-tips nearly touch the water. A flock of them is congregating about 500 meters ahead and slightly to port. The captain spots the birds and points the boat towards them; it might be that they have stopped there to catch small fish trapped at the surface by tuna.

We are fishing with lures that we hope will fool these very special fish into striking on what they believe to be a small squid – a method called trolling. I have my doubts as to whether it will work. The boat engines are making quite a noise and the lures are splashing about in the foamy wake of the vessel in a way that I fear will make it hard for hunting fish to catch them. But as we approach the birds, our reels suddenly scream into life. Line is torn out at a terrific rate as whatever has snatched the lures streaks off away from the boat. We have just gone over a shoal of feeding fish, and three of them have fallen for our fake squid. I pick up my rod and engage the reel to retrieve.

The resistance is instant and strong. I have hooked into an advanced ocean predator, and it has immense power and speed. I am used to fishing on little English rivers and lakes, where I never catch fish like this. I lift the rod, pull up the line and then let it drop. For a split second there is a little slack, and I quickly turn the reel to make the most of that, in shortening the line between me and fish by about 1 meter each time I do it.

Nye gets his fish back to the boat and onboard. It is about 3 kilograms and a very nice one, but as is befitting a father-and-son fishing trip, mine is bigger – and still fighting hard. Ricardo, one of the crew, is telling me to hurry up, in case my fish throws the hook and escapes. Eventually I can see it in the clear water. It is about 5 meters down, but now next to the boat. The line is tight and the fish is still resisting with power. I can see a blue torpedo shape tearing from side to side. It comes closer and Ricardo grabs the line. He tells me to reel in some more.

Eventually the fish is at the surface and he grabs the rubbery body of the lure in which the hook is embedded, and hoists the fish on to the little wooden fish landing deck attached to the stern. I politely ask if we should release it. He looks at me with incredulity, takes out a small baseball-bat-like implement and smashes the fish on the head. A single sharp blow is delivered. The fish quivers and then, as if a powerful energy source has been switched off, it becomes still.

Ricardo takes the fish by the tail and passes it to me. He has a look of pride on his face, rather like you might expect from a midwife handing over a new baby to her mother. My new arrival is a skipjack tuna, about 6 kilos. It is one of the most amazing creatures I have ever set eyes on. To call it streamlined does no justice to this highly evolved ocean hunter. Its body shape is exquisite; fins and tail are perfectly adapted for speed and trim. The solid body of the fish is packed with taut muscles that propel such animals at high speed after their prey.

It has huge eyes to help it see in the twilight world that lies beneath the immediate surface. On the top of the fish is an indescribably beautiful pattern comprised mainly from two sublime shades of aquamarine. Beneath and on its sides the fish is paler and silvery, so that, when viewed from either above or below in the shafts of light scattered from the waves and wavelets on the sea surface, it is apt to disappear. With this advantage it can surprise its prey, and also evade the sharks, dolphins and others that might make a meal of it.

This predator is by definition near the top of the food chain, but not quite. As the boat heads back toward port we see some of those higher still in the ocean energy pyramid, including bottlenose dolphins and short-finned pilot whales. The latter animals live in the waters between Tenerife and neighboring La Gomera, where about 300 of them feed on a population of giant squid that dwell in the deep water there.

The predatory fish and sea mammals that made our day are the most highly evolved and visible manifestations of the awesome productivity of the oceans. Even though the skipjack is the smallest of the commercially important tuna, these fast growing animals can get up to nearly 34 kilos. By the time such animals are in a small metal can on a supermarket shelf, it is perhaps more difficult to appreciate the processes that help make such great food. The rapid accumulation of protein in fish like this, not to mention the oils which are so beneficial for human health, depend on entire ecosystems to produce. And, as is the case on land, those ecosystems are ultimately powered by sunlight.

All around the world, drifting in the water at the surface of the oceans that cover two-thirds of our planet, are numberless tiny organisms that use sunlight to make complex molecules from inorganic materials. A variety of photosynthetic algae are the primary producers, using sunlight to create the materials they need to grow and reproduce. These microscopic light-powered life-forms that drift in the sparkling sunlit surface water are in turn food for a number of very small plant eating animals. These tiny creatures, floating along with the minuscule plants, include various single-celled amoebas, and also newly hatched fish, mollusks, jellyfish and crustaceans. These little animals are food for, among other things, the small fish, which feed the bigger fish, squid and others, and which in the end sustain some of the sea mammals and, indeed, a lot of humans, too.

By the time my 6-kilo tuna met its end, it was probably three to four years old. Depending on the size of the female that spawned it, that fish came from a batch that would have numbered between about 100,000 and 2 million eggs. Once released, skipjack eggs float at the surface of the sea, where they hatch in about a day. The tiny baby fish drift in the plankton, and thus become food for the many creatures feeding there, including slightly bigger skipjacks. The hatchlings that survive, though, grow quickly and after two to three weeks are fully formed juveniles that begin eating other small creatures, including fish, crustaceans and mollusks. The young adult skipjack soon becomes an opportunistic predator with a highly varied diet, and they grow rapidly, reaching about 40 centimeters by the end of their first year.

As a general rule, it takes about ten times the volume of food from one level in a food chain to grow an animal at the next level. Using this basic ecological principle we can guess that my 6 kilos of tuna would have required about 60 kilos of small fish to get that big. The 60 kilos of small fish would in turn have required about 600 kilos of animal plankton. That would have needed about 6,000 kilos (6 tonnes) of light-powered photosynthetic plankton.

When you start to multiply these kinds of numbers against a whole school of tuna, never mind the annual catch of countries like Spain, or the annual consumption of tuna loving nations such as Japan, it is evident that a mind-bogglingly broad pyramid of ocean production supports major industries, eating habits and restaurant profits.

The tuna are, of course, just one group of fish that are among a total annual wild sea fish catch which peaked in 1996 at about 86 million tonnes. Even if we assume all these fish are one level down from predatory tuna, that is over 8 billion tonnes of primary producers required to sustain such an output. With numbers like that in mind (and it is certainly an underestimate considering much of the fish we eat are predators like tuna, swordfish and cod), it is perhaps no surprise that about a third of the photosynthesis taking place on Earth is in the oceans.

Sometimes the top and bottom of the food chain are quite visibly connected in how catches rise and fall. Algal blooms can occur because of nutrients brought up from the bottom of the sea. As the microscopic plants rapidly increase in number, fed by the nutrients they need to grow, the great clouds drifting in the sea can sometimes be seen from space. This is generally received as good news by fishermen because such explosions in algae are often followed a couple of years later by an expansion in the size and number of fish that are caught, as energy and nutrients move through food webs to fuel fish growth.

Blooms in marine algae are not only produced as described in Chapter 2, through the runoff of nutrients from farms coming down to the sea in rivers, but also by natural processes. The most productive fishery on Earth in terms of the tonnage of fish landed is the anchovy fishery in the eastern Pacific Ocean off the coast of Peru. It is not just sunlight and photosynthetic algae that make this possible. At the bottom of the sea here, and in common with other fertile areas of coastal ocean, there are rich natural accumulations of nutrients.

The nutrients found on the deep seabed along the Pacific coast of Peru are made up from the decayed remains of dead animals and plants and the droppings of the animals living above. At the bottom of the sea, where it is dark and still, the buildup of organic material comes from a steady fall of "marine snow," gently coming down from above and the zone of life-giving light. As this organic material decays, it consumes virtually all of the oxygen dissolved in the water near to the seafloor. The conditions that result encourage the formation of fine muds rich in nutrients. These, in turn, foster various chemical processes, including those undertaken by bacteria that live in oxygen-free conditions, where they work on breaking down material such as the bones and scales of dead fish. From time to time, and because of factors that can include seasonal changes to the wind, these muds are disturbed and can be taken to the surface of the sea in so-called upwellings.

The arrival of these dissolved nutrients – especially nitrate and phosphate – into the upper part of the sea where the sunlight penetrates can fuel massive blooms of algae, which in turn broaden the base for the food pyramids founded upon them. The productivity of many fisheries is dependent on this kind of nutrient input. Whereas farmers apply fertilizers to improve pasture to increase the output of meat or milk, the recycled nutrients that support the marine equivalent of grass are applied for free, and in the process a lot of very healthy fish protein is produced.

These stocks of eastern Pacific anchovy yield around 7 million tonnes of fish every year, and in 2008 made export earnings for Peru worth about $1.7 billion. Much of the catch, however, has one further step in the food chain, as it has become a staple for cheap farmed fish, particularly salmon. The protein built by the anchovies is passed to the salmon in their feed, as are the healthy oils they make.

The value that humankind derives from ocean productivity through the fish we catch is truly vast, most obviously through nutrition. Official advice to eat fish is based on the excellent protein it supplies, as well as its health-giving oils and micronutrients. There is reason to believe that fish is important in neonatal development; one study found that pregnant women who eat no fish were more likely to have children with lower IQ levels than those who did. That 6-kilo skipjack fed five of us with fantastic meals for a week. Nye's one was very much appreciated by our Canarian neighbor and her family.

And, of course, in addition to amazingly good food, this ocean productivity is also the basis of economic activity and jobs. Fishing, fish processing and sales contribute an estimated $274 billion to global gross domestic product. These industries also deliver massive social benefit in the form of about 200 million jobs. The vast majority of this employment is located in developing countries, where jobs are often very scarce. And in many of the poorer parts of the world it is not only a question of fish providing healthy food – it is vital for nutrition in a much more fundamental way, for it is not only a healthy choice: with few other sources of protein, it is for hundreds of millions of people, the only choice.

The South China Sea

Some 120,000 fishing boats work from Vietnamese ports, river mouths and beaches. There are also various fast-expanding fish, lobster and shrimp farming enterprises. These industries between them support some 4 million workers.

Binh Dinh is one of Vietnam's twenty-eight coastal provinces, located in the center of the country facing the South China Sea. About 8,000 boats are based here, and about two-thirds of these are small craft supporting the families of inshore fishermen working the waters out to about 10 kilometers. The brightly painted boats are parked in shallow bays near beaches, or moored in sheltered palm fringed estuaries. Most of them are bottom trawlers that catch fish by dragging nets along the seabed.

On some of the sandy beaches, so-called lift nets are installed. These large and static fine-meshed devices are lowered from long poles so that they lie on the seabed during the night. They are lifted up in the morning, taking with them all of the (mostly small and juvenile) fish that happened to settle on them when it was dark. Many areas of the coast have also been developed for shrimp rearing ponds. From the air (and you can see them clearly on a Google map) they look like a mosaic of mirrored tiles arranged along the coast with pencil thin lines dividing them up. In the sea there are cages that house lobsters being reared for export. Various kinds of traps are in evidence, too, to catch different fish and crustaceans.

At the main port of Quy Nhon there is also a fleet of offshore vessels which fish further out to sea. These bigger craft can travel up to 100 kilometers from land in search of ocean-going species, including tuna, mackerel, squid and others that attract good money in export markets. Most boats fly the Vietnamese flag, with its single yellow star on red.

The overpowering smell of rotten fish is provoked into a nauseating stench by the constant warm temperatures. It is, however, the smell of success, and the people who bathe themselves in this aroma are earning the money they need to feed and house their families. The very successful boat owners are rich and own cars, while many of the regular fishermen are able to not only support their families but to have a motorcycle and mobile phone. In this fast developing country, both of these are ways to be more connected, to gain social mobility, earn more and thereby fuel further economic growth.

People are busy everywhere, preparing fishing gear, making minor repairs and loading supplies. A truck parked on the quayside is packed with 10 kilo blocks of ice. A mist gently falls around the truck as the damp tropical air comes into contact with material so incongruous with the local climate. The ice blocks are fed into a diesel-powered ice-crusher. Each time a new one enters the grinding jaws, the engine strains and a dense dark cloud of black smoke puffs out as it reduces the frosty blocks to splinters. The smashed ice is directed down a chute and into the hold of a trawler.

A couple of squid boats are in port. They look like floating junkyards. Piled on a platform above the main deck is a great clutter of poles and strings. These will be hoisted aloft when the craft is at sea so that its crew can dry their catch in the sun. There are about twenty little round bamboo coracle-like rafts stowed upside down on deck. These are launched at night to catch squid using hand-lines – a notoriously dangerous occupation. The rafts frequently drift away from the mother ship in the darkness, never to be seen again. However, the rewards and the absence of alternatives mean there is a ready supply of new recruits.

Small purse seiner vessels are tied up in port, too. Their crews are busy getting their nets ready for an imminent trip to sea. These craft catch fish that live in open ocean water, such as tuna. Their nets are laid around a shoal of fish and then the bottom is drawn together to create a big bag in which the fish become trapped.

A short drive from the main harbor a small shipyard is a hive of activity. Six longliners have been hauled out of the water for refitting and maintenance. The whine of circular saws is punctuated by regular hammering as new deck planks are fitted. The optimistic smell of freshly cut wood mixes with that of new paint. Two new vessels are also under construction, their thick bare skeletons of reddish wood taking shape from the keel upwards. The design is simple but solid, and the high quality tropical wood will serve well in resisting the various ocean creatures that will soon wish to make their home on the hull.

The width and quality of these beams is now hard to secure inside Vietnam, and such boatbuilding materials are mostly derived from the dwindling natural forests of Laos and Cambodia. As a timely reminder as to the consequences of continuing deforestation in the region, the very same week I stood in that boatyard at Quy Nhon, the critically endangered Vietnamese population of Javan rhinoceros was declared extinct. A casualty of massive deforestation (and poaching), it was a vivid example of how the liquidation of one resource (forests) was employed in the plunder of another (fish stocks), and how the depletion of both was destined to cause huge economic costs.

Van Công Việt is the owner of BĐ91251TS – a tuna longliner. His boat is about 15 meters long, displaces about 30 tonnes and can carry some 600 blocks of ice. He has been fishing for more than thirty years, the last ten specializing in tuna. His boat catches yellowfin and bigeye, and the odd swordfish. It is a good business and he is pleased with the living he makes. He takes his boat and crew of ten out for five or six thirty-day trips per year.

Việt's boat is designed for one purpose – catching fish – and human comforts are few. The crew's pillows and bedding are stored in the roof of the cabin, where the main table also serves as a platform for sleeping upon. There is a tiny galley for preparing meals, but no toilet. Two plastic beakers in holders on the wall of the cabin hold the crew's toothbrushes. There is no cabin space below deck. The rear half of the boat is taken up by its 180 horsepower diesel engine. The front portion has the ice and holds where the catch is stored. On the prow are two huge bamboo baskets, and in these are coils of nylon line. To these are fitted large, sharp, shiny steel hooks. Each long line is about 50 kilometers long and equipped with about 1,300 hooks. They are baited with squid and trailed behind the boat. A reeling device is fitted on the front of the boat to retrieve the lines after they have been deployed at sea – hopefully with a large number of tuna attached.

Việt is proud of what he has achieved, in recent years earning enough to buy a second boat and employ a second crew. But he is worried about the future. "During the last ten years the tuna stocks have fallen by about forty percent," he believes. He puts some of the blame on the purse seiner vessels, which catch smaller skipjack and also juvenile yellowfin and bigeye before they have spawned. But the expansion of the longliner fleet is a huge factor. One veteran of the

Binh Dinh fishing scene told me that in the late 1970s there were just five such boats operating here. Now there are more than 600.

Another longliner captain tells me he could once catch enough fish to make his living through trips that lasted five to seven days. Now he needs to go to sea for thirty days at a time. This means he has to buy more fuel, and that his fish is not of such good quality by the time he comes back, reducing its value by up to 70 percent. With such a cut in income and increase in expenses, he feels economic pressure to fish more. The longliner fishermen are caught in a spiral whereby the worsening plunder of the resource encourages further depletion.

It is not only the high seas fleet that is experiencing a squeeze between the number of boats and the fish available to be caught. The small-scale inshore operators have the same problem – only worse. One vessel owner tells me that stocks are declining and some species have been gone for several years. He reckons that in ten years the inshore fishery will be finished. "What will people do after that, where will they go to make a living?" he asks.

Because the most valuable species have been fished out, much of the catch landed by the inshore fleet is what is called "trash fish." This term applies to all the fish that cannot be eaten by people – though it can be eaten by pigs, chickens, farmed catfish, lobsters and shrimps, and that is where much of it now goes. Reliable data is hard to come by, but perhaps as much as a quarter of the fish caught in Asia heads in this direction. Ultimately, many of us in the West eat the products of trash fish, in the form of farmed tiger and king prawns and catfish (sometimes called river cobbler), as well as chicken and pork.

No one knows what the different sources of "trash fish" are comprised of, but it includes many juveniles from what would otherwise be commercially valuable kinds of fish, should they have made it to adulthood. And some of the fish being fed to prawns and lobsters are believed to belong to species that have been listed by conservationists as in danger of extinction. Rare or not, if young fish are killed off in this way, then the long-term implications for fish stocks appear all too clear.

In Binh Dinh, it seems that every conceivable method to harvest the riches of the productive seas has been considered, tried, refined and adopted. Some 46,000 people are believed to be employed there directly in fishing, with further jobs generated in processing, sales and boatbuilding. And, of course, the success of the fishing industry – generating a quarter of the province's GDP – has a knock-on effect on the whole regional economy. The Yamaha and Honda mopeds that swarm the streets, and the vast number of LG, Nokia and other mobile phones sold here, are in no small part affordable because of the productivity of the ocean.

In recent years, we have become more adept in understanding the economic benefits to be taken from wild fish. We have, however, been less skilled at understanding how the wild marine creatures that have such a huge financial value are derived from natural systems that need to be maintained in order to sustain their productivity. It is not only through nutrient-fueled algae growth that ecosystems produce the silvery creatures we depend on for so much. Natural marine systems work in a wide range of other ways to support wild fish populations.

The people who often appreciate this basic point most are the people who actually go fishing. Some of the fishers of Binh Dinh, for example, told me how particular ecosystem features sustain their livelihoods. One young man expressed his concern about damage being caused to areas of reef by trawlers. "The coral reefs in the inshore areas are nurseries for the young fish, but still people fish there," he lamented. "Education is needed."

It is not, however, solely in the developing countries that greater awareness of the consequences of degrading natural marine systems might be helpful.

Sea Forests

The long leathery seaweeds known as kelp are another kind of algae, at the other end of the size spectrum to the tiny single-celled plants that drift at the surface of the oceans. These giants are firmly attached to the seabed and, in the case of the California giant kelp, can grow up to 30 meters in length. Gas-filled chambers keep the long fronds afloat and near to the light that they need to grow.

Where there is a rocky seabed on which they can attach themselves, the kelp can grow into dense stands quite understandably referred to as forests. And like forests on land, the kelp forests in the sea are of vital importance to a wide range of different animals, including many species of fish, and among them the young of commercially important species that use the forests as nurseries. One factor that determines the presence (or absence) of these habitats is, surprisingly, a mammal – the sea otter. These are the only sea mammals that don't need a layer of blubber to keep warm. Instead they have an incredibly dense fur – an excellent adaptation for survival that paradoxically nearly led to their extinction. Sea otters once lived around the coasts of the whole North Pacific, from northern Japan to California, but they were hunted to the edge of oblivion during the eighteenth and nineteenth centuries, so as to supply a booming fur trade. Protection introduced during the twentieth century allowed some populations to slowly recover, including along parts of the coasts of California and British Columbia.

Despite their thick coats, sea otters still need to eat a lot so as to fuel the intense metabolism that keeps them warm, and each day they consume sea urchins, crabs and other items weighing around a quarter of their body weight. When sea otters were nearly made extinct, their voracious appetites were removed from the kelp forests. As a result the population of the sea urchins, which formed a major part of their diet, exploded in number. In extreme cases the removal of sea otters led to so-called "urchin barrens" – areas of seabed grazed to almost bare rock. The urchins' powerful jaws and strong teeth munch through the fibrous stems and leaves of kelp, and also invertebrates attached to the seabed, such as sponges and fan worms. And when the kelp forests went, so did an ecosystem where young fish grew up.

Urchins are not just a problem in the Americas. On my trip to Tenerife I took my other son, Sam, on a diving expedition with a company called Ocean Dreams Factory. As well as taking tourists into the water to see wildlife, David Novillo (who founded the company) is also waging a vigorous campaign to control urchins. The urchins have laid waste to the seabed around the island because the coastal fish that once kept their numbers down have been so comprehensively depleted by overfishing. He and his team of divers spend a lot of their time removing urchins, and in one bay where they have worked especially hard the underwater ecosystem is coming back. Where the seagrass beds are recovering, green turtles come to feed. Sam and I swam with one.

While the recovery of kelp forests and seagrass beds give some encouragement that changes to natural marine systems can be repaired, a note of caution is warranted. Sometimes, and as is the case on land, the changes we make to ecosystems are not always simple to reverse. Indeed, it may be that, once altered, some will not return to their original state, even when the cause for the change has been completely removed. A case in point is the Grand Banks cod fishery off the coast of Newfoundland.

The cod populations here were legendary. An account from 1497 reporting on the voyage of John Cabot spoke of a sea "swarming with fish that can be taken not only with a net but in baskets let down with a stone, so that it sinks in the water." Other accounts told of cod as big as a man. Not surprisingly, this huge and productive population was for centuries targeted with ever more effective means of capture. But in 1992, the bonanza came to an abrupt end as the stocks that had been fished for hundreds of years fell into a state of crisis. A ban on cod fishing was introduced, and it remains in force today.

Why after twenty years the fish have still not come back is, as yet, unclear. What is evident, however, is that a change in the marine system has occurred, and that this is for now preventing a bounce back in the cod. The dynamics that once existed between the cod and the system that sustained them in such vast numbers have altered. Some believe a high population of harp seals prevents their recovery, while others point to a possible link with an explosion in lobster numbers. These crustaceans were once eaten by the cod, but when the cod numbers fell the lobsters increased and prevented the recovery of the cod through eating the fishes' eggs.

Whatever the reason for the cod not yet coming back, it is important to remember that, when the fish went, some 20,000 jobs went with them, and so did a major part of the economy of Newfoundland. And in this case it is not simply a question of waiting for stocks to recover: it may not happen.

Today, about a third of fisheries across the globe are being exploited beyond their maximum yield. Some have already collapsed and perhaps half the total are at the limit of what they can provide; many of these lack effective regulation and management, so may soon become overexploited. In both cases, matters are made more difficult by illegal and unregulated fishing.

Professor Callum Roberts, a marine biologist based at the University of York, spends a lot of time thinking about the state of the oceans and our impact on them. He explained to me the historic process: "In the early twentieth century the footprint of fishing spread from traditional fishing grounds to distant seas, and then in the late twentieth century from coasts to the high seas, and shallow water to deep. Over time we have substituted new species as past favorites waned. The price of fish has outstripped inflation for decades, reflecting the increasing difficulty and cost of sustaining supplies. In the last third of the twentieth century, developed countries turned to the developing world to supply their fish, having exhausted their own grounds."

Because of rising human numbers and a growing demand for protein, Roberts sees little prospect that the pressure to catch fish will go down any time soon. And this will have consequences: "If we exploit the oceans as we have done for the last century, then overfishing will continue to eat away at populations of the world's big fish; some will be driven to extinction while many more will become too rare to play any further meaningful role within their ecosystems. As they disappear, we will continue the switch from large predators like cod and hake to animals low in food webs, like prawns and anchovies. But they too will become overexploited, as some already are, and we will have to seek seafood from other sources, such as Antarctic krill, which will be processed into more palatable looking foods, like fishcakes and fish sticks."

In the face of the progressive collapse of wild marine fisheries, some have advocated an expansion of fish farming as the logical answer. And this has indeed been part of the response – to the point today where the amount of farmed fish consumed exceeds that taken from the wild. But this has not ended our dependence on ocean systems, not least because much of the farmed fish is fed small sea fish, taken from the wild.

Progress is being made in bringing forward partially vegetarian diets for some farmed fish, including the catfish being grown in Vietnam and salmon in Scotland, but the link with the sea remains. It is also important to remember that farmed fish need to be tended and treated, which costs money. The wild fish replenish themselves for nothing – those light powered food webs contributing hundreds of billions of dollars of GDP, jobs and food for no investment (save the fuel and equipment to catch the fish).

Fish Forever

One impetus for a more determined effort to sustain the wild populations of sea fish has come from a recent World Bank report called Sunken Billions. This concluded that an extra $50 billion worth of value could be added to fish stocks each year if only they were better managed.

And there are some encouraging signs – places where overfishing and decline have been met head on, and increased economic value actually achieved. Each place is different and there are many tools and methods that can be used to improve the long-term productivity of fisheries. But once the will to make change is there, amazing things can be quickly achieved.

One example is the halibut fishery in the North Pacific, where a reform program that cost the equivalent of about 3 percent of the annual revenue earned by the industry was spent in ways that increased its productivity. Income from catching this valuable species increased from around US$50 million a year to $245 million – an improvement of 390 percent, and by any standard a good return on investment.

In New Zealand, some US$25 million was spent on better fisheries management, and as a result the national value of fisheries increased from $1.57 to $2.3 billion – an increase of 46 percent. Norway spent about $90 million in reforming its fisheries, including putting in place a ban on discarding any fish that had been caught. This helped change fishing practices, stocks recovered, and the value of the annual catch went from $347 to $546 million.

Analysis of the potential economic gains that could be derived from better managed fish stocks suggests that there are many such opportunities. For instance, getting the disastrously managed Northeast Atlantic bluefin tuna fishery into better shape could lead to gains of up to $510 million per year.

In the developing countries, too, there is cause for optimism based on what a few pioneers have achieved. Namibia has introduced an effective vessel monitoring scheme that has dramatically reduced illegal fishing and enabled stocks to recover. Catches have increased threefold; so has the benefit to Namibia's development, with the value of fishing to the national economy going up from $98 to $372 million per year.

In Vietnam a modest level of investment enabled a community management scheme to be set up by local people at a clam fishery in Ben Tre Province. After making the transition to better management of the resource, the clams now support 13,000 households, compared to 9,000 in 2007.

As is the case in the richer parts of the world, there is a growing body of analysis pointing to the scale of the economic opportunity that could come with improved fish stocks. For instance, one study suggests that the Hilsa Shad fishery in Bangladesh could be worth nearly $260 million more annually if it was subject to better stewardship.

These and other positive cases of fisheries reform have been backed by various economic incentives; for example, granting fishermen property rights, thus helping end the "tragedy of the commons" and the situation in which no one had an incentive to conserve the resource. With rights in place, they no longer had incentives to catch fish before the next boat did. The result was less plunder and less bad practice.

There is plenty of scope for mobilizing more finance for better fisheries, not least through changing how subsidies work. In many parts of the world public money is being used to promote over-fishing, for example, by providing the funding that has caused fishing fleets to grow too large. Subsidies that damage fish stocks are estimated to total about $16 billion per year globally. Develo-ped countries spend twice as much on these types of subsidies than they do on protecting the oceans. Part of the answer here is to change the economics so that activities which sustain Nature and its productivity get rewarded, while those that cause unneces-sary damage have support progressively withdrawn.

Some of the fish stocks that have adopted better practices have been assisted by certification and labeling schemes. The Marine Stewardship Council certifies fish that meet certain minimum standards. Some of the tuna fishers I met in Vietnam would like to certify their catch. They have a lot of work to do, and a lot of changes to make, but they seem willing to go on the journey. Long-line captain Van Công Viêt, who we met earlier, was already in the process of replacing J shaped hooks with special C shaped ones that prevent the accidental capture of endangered sea turtles.

Tuna fishing, too, is showing signs of better regulation. The Inter-national Sustainable Seafood Foundation has encouraged processors only to buy tuna from identified vessels. This has helped to stop ille-gal fish entering the market and has given an economic boost to those vessels operating within the law.

The tuna business needs this and other measures, for there is big money involved. In 2012, a single bluefin tuna was sold at a

Tokyo fish market for nearly three-quarters of a million dollars – $2,737 a kilo – nearly doubling the previous record price paid for such an animal in 2011.

This species of tuna is especially at risk, with some populations already gone and others near to collapse. This particular specimen was a monster, weighing in at 269 kilograms, and was bought by an upmarket restaurant and sushi chain. The price was a reflection on the increasing rarity of this species of tuna. However, if it is possible to use the high prices paid for some species to create incentives for sustainable management, then perhaps even for these sought after ocean predators there can be a long-term future.

It is perhaps natural that much of the (limited) attention devoted to the conservation of what the oceans do for us has in recent years been devoted to the conservation of fish. After all, we eat them and they have a very tangible economic value. In the greater scheme of things, however, the seas do a lot more than supply fish. And the economic value of those other benefits is truly huge.

Trillions of tiny coccolithophores form a plume off the coast of Norway.

CHAPTER 8
OCEAN PLANET

$21 Trillion: Annual economic value provided by the oceans

Over 50 Percent: Proportion of oxygen produced by plankton

99 Percent: Proportion of planetary living space that is in the sea

FOR US AIR-BREATHING, land dwelling animals it is natural to see the world first and foremost from the perspective of terrestrial environments. But, as any child will tell you, most of the surface of the Earth is covered with salt water. These two worlds, one defined by water and the other by air, are fundamentally different in character. Whereas on land, Nature mainly works from the soils and the bottom-up, the ocean system runs from the top-down. The complex food webs that maintain it are based on multitudes of larger creatures feasting on the products of the solar-powered productivity in the well lit surface layers.

Water is much denser than air and one consequence of that is how it supports the weight of the creatures that live in it. In the thinness of the air just a few species of birds, such as swifts, feed on the sparse "plankton" of aphids and other tiny insects that are wafted aloft. In seawater, animals live not only in the top layer, where most of the productivity takes place, but in all the layers that go way down to the bottom. And across much of the oceans that place is very far indeed from the surface, in some cases much further than it is from sea level to the tops of the highest mountains. Not many people have ventured into the dark world beneath the waves, to the cold and alien space beneath. One who has is Sebastian Troëng.

Troëng is Swedish and grew up on the shores of the Baltic Sea. His fascination with the oceans goes back to his early childhood and carried him into a career as a marine biologist. He has had many roles with different organizations and today works with Conservation International, where he leads a global program aimed at the intelligent use of the oceans. One of his organization's supporters is based in the Bay Islands of Honduras, from where he operates his own submarine. He built it himself, and now makes deep dives off the coast of Ruatán. This island lies next to deep water and minimal currents, so it is a good place to take a submarine to sea.

Troëng traveled in this little vessel to the zone far below the surface, where no light penetrates and where the crushing pressure of the water would kill air-breathing animals in an instant. He told me what it was like.

"It was hot and steamy, about thirty degrees, as we were towed out by a tourist boat. When we were in position we closed the lid and started our dive. We began a slow and very gentle submergence. The first light to go is the red light, and at about 100 feet everything that is red appears black. There is less and less light, and by about 300 feet it started to look like dusk. By the time we reached 700 feet, it was total darkness. At about 2,000 feet we found the bottom. As you go down it gets cooler and cooler. I tried not to touch the side, because it was dripping with condensation. It's not recommended if you suffer from claustrophobia."

"We drove the submarine around the bottom in places that in all probability had never been observed by humans ever before. There were translucent fish, six-feet-long phosphorescent animals, huge jellyfish and deep sea corals."

"It was a very varied seascape. Some areas were sandy and covered with urchins and a few lobsters running around. And then we came to a rocky area where there was a sea fan. Given the limited nutrients down there, these animals grow very slowly indeed. Some of them would have been 500 years old. They started growing around the time when Columbus arrived in the Caribbean. There were starfish, little fish, and crabs, all living on the teeming ecosystem of a sea fan."

"2,000 feet feels pretty deep, but it is only about a sixth of the average depth of the Earth's oceans. But going down there made me realize that most of the livable space on the planet actually looks like this. Pitch black and at extreme pressure and most people will never see it."

Troëng reminded me that, with immense depth and covering most of the surface of the planet, the oceans represent about 99 percent of the living space on Earth. And in this huge system multiple complex processes are underway, including those that affect the Earth's carbon cycles.

The Oceanic Carbon Sponge

The coccolithophores which we came across in Chapter 6, the tiny organisms which release the dimethyl sulfide that produces the sulfur particles that help cloud formation, also play a vital part in the Earth's carbon cycles. These tiny single-celled algae are enclosed inside an intricate calcium carbonate cage comprised of little discs that look like hubcaps.

Although they are so small that to see them clearly requires an electron microscope, their numbers are so huge as to make a difference to the entire world. When carbon is locked up in these delicate shells it means there is less of it to unite with oxygen molecules to form the main greenhouse gas – carbon dioxide – in the atmosphere.

When they die they sink to the seabed, taking the locked up carbon with them. Over millions of years, accumulations have built up on the seabed to form layers of what is now chalk and limestone. This has been a major factor shaping the past climate of the Earth, and remains important today. About half of the CO_2 taken up annually by photosynthesis is via organisms living in the oceans, including the coccolithophores.

These and other organisms could be at risk, however, because the oceans are turning more acidic due to increased concentration of carbon dioxide in the atmosphere. With more CO_2 in the air, more is dissolved in seawater, leading to an increase of carbonic acid in the marine environment. Dr. Carol Turley, a leading marine scientist based at the Plymouth Marine Laboratory in southwest England, is among a growing number of ocean researchers who are increasingly alarmed at this progressive acidification of the seas, and describes ocean acidification as "the other carbon dioxide problem." She points out that since the start of the industrial revolution, over 200 years ago, the oceans have already taken up nearly 30 percent of the CO_2 we have liberated.

"This is making the oceans more acidic and affecting different steps in the carbonate cycle in the oceans, including the carbonate

ion concentration, and if this continues, it will affect the shell production of animals like mussels and oysters, but also the formation
of coral reefs. Increasing acidity also has an effect on the physiology
of organisms. As we put more CO_2 into the atmosphere, then ocean
acidification will increase and waters will become more corrosive
to unprotected animal shells."

"It's happening globally, but because cold water absorbs more
carbon dioxide than warm water, the acidification effect is expected
earlier in the Arctic and Antarctic, with major changes to the chem-
istry in just a decade or two. By the middle of the present century
there could also be serious ecological impacts in the tropics too,
and it may reach a point where the growth of coral reefs may be
lower than the rate of natural erosion affecting them. People alive
today are likely to witness these changes." A remarkable fact about
this kind of profound change to the Earth system is the infrequency
with which this level of shift is believed to have taken place in the
past. Turley says that Paleo-oceanographers believe that the last time
a similar event took place was about 55 million years ago. Back then
the progressive acidification occurred over several thousand years,
not in a few hundred, as is the case now. "What we are doing today
is a very rare event on planet Earth," she says. "But of course we can
stop it." Turley's research has not only highlighted the effects of
acidification but also two other factors that are causing potentially
profound changes. One is the progressive heating of the surface of
our oceans and seas because of the increase in average global temperatures, also caused by the buildup of greenhouse gases: "When
you heat the surface of a liquid you can stratify it, create layers. It's
like when you go for a swim in the sea in the summer and you find
the warm bit is on top and the cold bit underneath. When you get
layers like that, then you can get less transfer of nutrients from the
colder deeper water and that in turn can affect ocean productivity.

This in turn can have a knock-on effect on food webs and fisheries."

The impact of this is perhaps already evident, for example in the recent collapse of the sardine population in the southern Caribbean. In this case, the drop in the catch from about 200,000 tonnes in 2004 to about 40,000 tonnes now has been linked to a decline in plankton, in turn caused by a cut in the nutrients reaching surface waters, with that brought about by warming of the sea.

The other change, and which is directly related to the heating of the sea surface, is the reduced level of oxygen: "As you warm liquids, gases are released and in this case that means less oxygen dissolved in the seas and, of course, since most of the animals need oxygen, this can cause problems for life in the sea if too much is lost."

As with so many ecological changes, Turley points out, a major concern is that these trends will not only cause changes on their own, but in concert with one another: "While scientists have been focusing on one or other of these stresses, many areas of ocean are suffering from two or three of them at the same time. The impacts of all three can cause major changes. Unfortunately the areas where the hotspots occur, where all three factors are having a major impact, also tend to be the areas of highest productivity."

When it comes to the coccolithophores, experiments suggest these organisms are quite sensitive to ocean acidification, although some strains may be resilient to slightly more acidic conditions. But even if they are, ocean acidification is a fundamental ecological shift that could lead to unforeseen changes in the oceans, and the wider Earth system. Irrespective of the effects of greenhouse gases on the Earth's climate, avoiding the acidification, layering and deoxygenation of the oceans are all very good reasons for us to invest in rapid cuts in carbon dioxide emissions.

As well as the myriad tiny plants drifting in the sea, massive carbon extractors and stores exist in different kinds of coastal marine systems, including the tangled growths of mangrove forests that thrive in shallow tidal tropical waters, salt marshes and seagrass beds. Even though these ecosystems cover just 0.5 percent of the ocean, they rank among the most intensive natural pumps on the entire planet for removing CO_2 from the atmosphere. These habitats are also important fish nurseries, and are vital in many parts of the world for coastal protection.

But they are also being lost faster than just about any other ecosystems on Earth. The coast of Vietnam provides a case in point, as natural areas have been cleared and turned over to activities that provide a more immediate and visible economic return. And they are far from alone. North America and Europe changed much of their natural coastal areas some decades or even centuries ago, and the rest of the world is catching up fast. Some kinds of coastal ecosystem are being removed at a rate of up to 7 percent per year, and at that rate most will be gone within a couple of decades.

If, on the other hand, we take effective action to stop the decline, and to begin restoring these carbon-heavy systems (so-called "blue carbon"), this would account for about a tenth of what we need to do in order to stabilize CO_2 concentrations in the atmosphere. That is a huge amount, and underlines the vast economic value of coastal areas – places that have for far too long been seen as unproductive wastelands, with more value when converted to ports, claimed for farming, turned over to shrimp production or cleared away to provide fuel.

And as carbon dioxide is absorbed by tiny sea plants and made into carbohydrates and more complex compounds, so oxygen is released as a by-product. Take another breath. At least half the oxygen molecules you just sucked in were provided free by tiny marine life-forms, including the ones that help make it rain. The tiny organisms drifting in the sea help to not only keep concentrations of carbon dioxide down, but also the oxygen up. It's not a simple task to calculate just how much of the oxygen in the air is put there by the ocean plant plankton, but estimates range between 50 and 85 percent.

The fundamental planetary services provided by the oceans are evidently of huge economic importance. One study suggests that, of the total value to the human economy provided by Nature, some 60 to 70 percent is derived from the seas, with most of that linked to the services provided by coastal systems.

Taking a more specific estimate that the oceans are worth about 63 percent of the economic benefit we get from Nature, and applying that to one widely quoted estimate of the total value of natural systems developed by Oregon University Professor Robert Costanza in a groundbreaking 1997 study, the oceans can be calculated to be worth about $21 trillion per year.

Like so many of these ballpark numbers, Costanza's calculations have been questioned. But the specific numbers are not really the main thing. The important point is that there is huge value, seen in the fact that pretty much all of the rain that falls on land comes from the oceans, that they absorb about a third of our carbon dioxide emissions and produce much of the oxygen that enables life to flourish. Not to mention the fish.

Tentacled Horsemen

The extent to which we have yet to appreciate our reliance on the oceans is also seen in how we continue to treat these priceless systems as a rubbish dump. A vivid example can be witnessed far away from land, way out in the North Pacific Ocean. Sailing into the central part of this vast area of water reveals an expanse of sea that is rather like a plastic soup, now about twice the size of Texas – and growing.

Held in place by a vortex of ocean currents, the area of plastic now extends from about 900 kilometers west from the coast of California past Hawaii and almost as far as Japan. It is estimated that some 100 million tonnes of debris are now floating there. About a fifth of it is believed to have been chucked off ships and oil platforms; the rest came from the land. Toy bricks, footballs, yoghurt pots, kayaks, bits of cars, bags, bottles and all the other plastic paraphernalia that characterize modern life are floating around in the middle of the ocean. There are several such agglomerations of floating plastic detritus. Another one is awash in the Sargasso Sea, while another huge one has built up in the Bay of Bengal.

These great rafts of virtually indestructible waste are powerful reminders that sea is downhill from everywhere and, as more and more plastic is manufactured, sold and disposed of, so more and more finds its way out there to the middle of the remote ocean.

Jo Royle is a sailor and captain. She has spent her entire life connected to the sea and, although only in her early thirties, she has spent many years far from land, taking part in ocean sail races and as a member of oceangoing expeditions. She is acutely aware of how much plastic is building up in marine systems, and how the links between what we do on land fundamentally shapes what happens in the sea.

"There'd be times at sea," she told me, "when I'd sail through little patches of plastic waste and think, 'Oh my goodness, a big ship must have just passed here and dumped a load of rubbish.' It was quite distressing to see once pristine ocean, hundreds of miles from land, polluted in that way. I soon learned that it was too much to be from a ship, and that it was coming from the land."

"When you see people in the supermarket in London with plastic bags blowing around the car park in the wind, you know some of them will end up in the storm drains, and then in the river. After that the bags go in the sea, where they will join the great mass of plastic debris that is already floating there. Each ocean has its own currents which sweep past the land masses, and these pick up rubbish from around the coastlines. It's a vortex, like a toilet, but one that never flushes. It's pulling the plastic round and round and round, and it has nowhere to go. With plastic there is no 'away.'"

Far from land, this seaborne plastic is for most of us out of sight and out of mind. But unfortunately it has serious and widespread consequences, most obviously for seabirds and turtles that eat plastic items believing them to be food. A 2006 study from the United Nations Environment Program estimated that about 1 million seabirds die this way each year, and about 100,000 sea mammals. The same research reckoned that each square mile of ocean has nearly 50,000 pieces of floating plastic.

Even more troubling than its effects on wildlife is what happens to the plastic as it gradually breaks down, as those 50,000 big pieces per square kilometer turn into literally billions of smaller ones. When it degrades into tiny particles, mostly because of sunlight, plastic does not go away: it breaks into smaller and smaller pieces of the same material, rather like a rock weathering into smaller pieces of sand. Some of it is taken up by animals, including microscopic creatures drifting in the plankton. Once plastic has entered this level in food webs, it then goes higher, as smaller animals are eaten by larger ones. Over time the tiny particles – which have built up in the oceans like a time bomb – will enter the plankton and other marine food webs. There is major cause for concern in this, not least because the plastic "sand" attracts chemicals that are ingested by animals.

Millions of plastic specks have a greater combined surface area than one big one, and these little pieces are poison magnets, attracting seaborne industrial pollutants and chemicals that include substances such as PCBs and DDT. Toxic chemicals like these don't like to dissolve in water and prefer to combine with oily substances, such as plastic. The toxics hold on to the plastic and in the process become concentrated, by as much as 1 million times more than is found in the surrounding sea. Once something eats such contaminated particles the chemicals can get transferred into its body. This chemical build-up has already been noted as altering the hormonal system of swordfish – and, of course, we are the next step up in the same food chain.

There may also be consequences for the plankton that is so fundamental to the ocean food webs and carbon cycle. In these and other ways the plastics we are dumping with such abandon into the oceans could seriously undermine what the oceans do for us, and that in turn could come with a very substantial economic cost.

And it is of course not only plastics that we dispose of in the sea, but wastes running off the land, from agricultural chemicals and sewage. These have contributed to the so-called dead zones described briefly in Chapter 2.

With a multi-pronged assault comprised of overfishing, pollution, acidification and climate change, it is perhaps not surprising that indications of profound change taking place in marine systems are now evident. Jo Royle told me that one thing she has noticed as a sailor is the ever increasing number of jellyfish. "In some places it's really striking," she said, "for example, in the Mediterranean and around the north coast of England."

With fewer fish and other predators in the sea to eat small jellyfish, more of them survive to become big jellyfish that then lay a lot of eggs. This in turn causes changes to food webs, as the jellyfish devour huge quantities of plankton, thus depriving small fish of their food source. The jellies also eat fish eggs, so that once fish numbers are reduced it may not be easy for them to recover. The process is aided by warming seas.

One study found that some 2,000 species of jellyfish are appearing earlier in the year. In addition, they are able to reproduce faster and some tropical species are extending their range. And, while the low oxygen conditions caused by pollution entering the sea via rivers harms fish, jellyfish are able to thrive.

This causes not only inconvenience but economic damage, particularly for tourism. If the sea cannot be used safely as a result of too many jellyfish, then this can cause serious problems, including in areas already suffering from reduced trade because of other factors.

And there are other impacts from jellyfish infestations. On the Mediterranean coast of Israel north of Tel Aviv, there is a large coal-fired power station. Unlike some other such plants, it doesn't have giant concrete cooling towers, and this is because it draws in seawater to cool down. From its tall exhaust stack the gases released from coal combustion rise in a thin yellowish brown column. The heat generated boils the water that creates the steam that drives the turbines making the electricity to power Israel's ever more energy hungry cities. The coal-fired power station also produces the electricity that powers the desalination plant located right next door.

Both these pieces of infrastructure are vital for Israel's fast growing economy, but changes in the sea recently caused them to falter. I walked on the beach near there with Dr. Ruth Yahel, the chief marine ecologist with the Israel Nature and Parks Authority. She told me that local seawater temperatures along this coast have increased by about 4°C in just thirty years. Overfishing had already drastically reduced the number of marine predators, and the build-up of pollution and nutrients has increased as agriculture has intensified.

Into this fast changing ecosystem non-native species of

jellyfish have arrived. They came through the nearby Suez Canal from the Red Sea, which in turn is linked to the ecologically separate Indian Ocean biome. The alien jellyfish, entering a novel environment, exploded in number. One swarm was about 100 kilometers long and 2 kilometers wide. Thousands of them built up in the inlet pipes of both the power station and the desalination plant, causing a serious crisis.

A similar problem happened in nutrient rich waters off the coast of Japan, where jellyfish can grow to the size of refrigerators. In 2006, a mass of the animals became clogged in the cooling system of a nuclear power plant, forcing it to cut back electricity production.

The jellyfish first appeared in the Ediacaran era, at a time when life on Earth was simpler and less diverse than today. These resilient creatures have seen and survived five mass extinctions, and at the start of what many scientists are calling the sixth extinction – our own age – their numbers have grown, and in some cases exploded.

The faint impressions in the Precambrian rocks that I saw at Brachina Gorge – in Australia's Flinders Ranges – told of a time when such animals were among a relatively small handful of early multicelled creatures. Those fossils and their living descendants perhaps offer a warning that life is not static, that it changes and that ecological consequences arise from pressures placed on natural systems. It might be that today we are helping to create conditions more like those when these creatures first appeared; a simpler world with fewer species. Perhaps we should take the recent rise of these ancient creatures as a warning, the great swarms of jellies as tentacled horsemen heralding an ecological apocalypse? I sincerely hope not, but it seems the height of folly to ignore what is surely a sign of profound and rapid change taking place in the Earth's oceans.

The 600-Year Legacy

Whatever we choose to learn from the trends now seen in marine systems, it seems fair to say that efforts to conserve the oceans are still in their infancy. One indication of the relative neglect of ocean ecosystems is the fact that, while some 13 percent of the land is now in some form of national park or Nature reserve, less than 1 percent of the marine environment is safeguarded this way, and nearly all of what is protected is in coastal waters. This is beginning to change, but even if international targets are met and some 10 percent of the oceans are protected by 2020, this will still not address many of the systemic changes that are taking place.

If the benefits we gain from the oceans are to be maintained, then (among other things) remaining natural stretches of coast must be kept intact and others restored, carbon dioxide emissions reduced, land-based pollution (including nutrients from farming) cut back and plastic pollution dramatically reduced. If we had ways to better reflect the vast economic value of the marine environment, then perhaps all this might be easier.

Whichever way you look at this set of challenges, however, it is clear that some quite fundamental shifts are needed, including in the design, use and disposal of consumer goods. Jo Royle pointed out to me the case of a plastic drinks bottle: "It might be in actual use for about two minutes, and yet could spend upwards of 600 years gradually breaking down in the sea." She picks up my iPhone and points out how there are seventeen different kinds of plastic in it, making the product really hard to recycle. "I don't think it's a question of stopping the use of plastic, but more a matter of how we use it."

As far as the progressive acidification and warming of the seas is concerned, there are also some rather straightforward conclusions. The first is to reduce the amount of fossil fuel we are burning. The second is to halt the destruction of the peat bogs, forests and other natural systems that are soaking up so much carbon, including the coastal marshes and mangroves that are such huge carbon pumps.

And when it comes to all these systems, there is another very good reason for us to see them as having a huge value. People living in New Orleans might tell you why.

A band of coral reef in Belize, backed by an area of mangrove forest.

Chapter 9
Insurance

$81 Billion: Damage caused by Hurricane Katrina in 2005

25 Percent: GDP of Belize reliant on corals and mangroves

$200,000–900,000: Value of 1 square kilometer of mangrove forest

September 1, 2005. I am lying on the grass in the late summer sunshine of a London park. Next to me is Barbara Stocking, the head of Oxfam. We are there because of our interest in climate change — and, more importantly, what to do about it. Barbara's organization was becoming engaged with the subject because the lives of poor people would be especially hard hit. I was there as director of Friends of the Earth, whose principal concerns were the ecological changes expected to accompany rapid climatic shifts.

In the park, underneath the huge wheel of the London Eye, we and hundreds of others arranged our bodies to form the shape of a spiral weather system — a hurricane. The reason was the launch of a new campaign called "Stop Climate Chaos." We had adopted a spiral storm symbol for the campaign logo. The idea was to highlight how more extreme weather was projected to accompany increased greenhouse gases in the atmosphere and, through awareness of this, to encourage cuts in climate changing pollution.

The campaign, and the event to mark its launch, had been planned for months. We had no idea that the hurricane symbol we had adopted and were mapping out on the grass would be so timely. Two days before our launch, Hurricane Katrina had made landfall on the coast of Mississippi and Louisiana, and had subjected the city of New Orleans to some of the most punishing treatment such a storm could dish out.

The destruction wreaked by Katrina is well known. A Category 5 hurricane, its storm surge (more than 8 meters above the norm) overwhelmed New Orleans' flood defenses, leading to the inundation of some 80 percent of New Orleans.

Incredible scenes followed. Thousands of people sheltered in the city's Superdome stadium; many were without food and water. Widespread looting and lawlessness sharpened the city's pain. The failure of the federal authorities to respond quickly or effectively underlined how, even in the USA – the world's largest economy – extreme weather can cause chaos and crisis.

It is generally accepted that the failure of the city's defensive levees was to blame for the severity of the disaster that unfolded. This was undoubtedly a vital factor, but as commentators sought to find fault with human engineering, and indeed with government agencies, they tended to ignore the vital role played by Nature, or more accurately roles that had not been performed by Nature.

One person who sought to look beyond the most obvious causes was Professor Hassan S. Mashriqui, an engineer based at Louisiana State University at Baton Rouge. He and a group of colleagues looked at the impact of Katrina alongside that of a second massive storm called Rita. This one hit the south coast of the USA just over three weeks later, but about 475 kilometers to the west of where Katrina wrought its worst impacts, at the Louisiana/Texas border.

The researchers analyzed the impacts of the two storms not only from the point of view of manmade defenses, but also the natural coastal features where they made landfall. The team carefully mapped the tracks of the storms and the high water marks they reached. Computer models were then employed to assess the effect of coastal wetlands in shaping the extent of the flooding that resulted.

In the mid-1920s, the coastal areas that lay in Katrina's path were comprised of salt marsh backed by freshwater swamp forests with tangled thickets of cypress trees. From then until the 1960s, various large-scale shipping channels were cut through the marshes. Not only were the wetlands fragmented, the incursion of salt water which was then able to penetrate the inner marshes killed the swamp forest. Prior to these channels being cut and then progressively dredged and enlarged, New Orleans was protected by a wetland buffer some 16 kilometers wide. Where Rita made landfall, by contrast, much of the original wetland was still intact.

As Katrina made its way across the Gulf of Mexico, pushing a bulge of water before it, the water level in the shipping channels rose up. The researchers simulated the flow of water as it reached the coast and found that most of the water that reached New Orleans came through the dredged channels, rather than across the wetlands. Water also flooded out of the channels and across the remaining coastal wetlands, reducing the beneficial effect they could have in cutting the power of the surge.

Even so, where wetlands were still in place, they did make a positive difference in protecting the city. Most of the places where the levees were breached or overtopped were where they did not have a wetland buffer on their seaward side. The research found that levees fronted by substantial wetlands on the sea side suffered little erosion damage, even if overtopped. The wetlands reduced the energy in the waves, the height of the surge and the power of the currents as they tore at the levees. If the wetlands had been intact, they would certainly have made a difference to the impact of Katrina.

Rita, like Katrina, was a Category 5 hurricane – and the fourth most severe ever recorded in the Gulf of Mexico. Unlike Katrina, it didn't hit a city the size of New Orleans, and where it came ashore the wetlands were still largely intact. The difference this made is in part reflected in the fact that, while Katrina took more than 1,600 lives, Rita killed seven. Mashriqui and his colleagues concluded that coastal wetlands work as "horizontal levees" and can reduce the height of a storm surge and cut down the amount of energy in waves before they reach engineered levees and populated areas.

Sinking Land and Rising Seas

In Louisiana there is another reason why it makes sense to protect wetlands for the disaster prevention benefits: the land is sinking. There are two ways to respond to this challenge, one is to build levees higher, the other to let the coastal wetlands naturally build up, gathering sediments from the sea and rivers, compensating for the underlying drop in the land. The research by Mashriqui and his colleagues suggests that the second option would work better and cost a great deal less money.

About a fifth of the world's human population lives within 30 kilometers of the sea. As coastal systems are degraded and replaced so the vulnerability of coastal societies to sea level rise and extreme weather events increase. While in countries like the USA, natural features such as wetlands have been shown to have a major economic value as part of the coastal defenses, in developing countries where there is less money available for engineered protection in the form of sea walls and levees, they can be more important still.

Recent history is littered with examples of how coastal communities in developing countries are especially vulnerable to the impacts of extreme conditions. Perhaps the worst in modern times was the Bhola cyclone that hit East Pakistan (now Bangladesh) in 1970. This storm created a massive surge that flooded many of the low-lying islands in the Ganges Delta, in the process killing an estimated half-million people.

It is not only in relation to extreme weather that natural systems can provide protection and resilience in the face of sudden shocks. A powerful example came on Boxing Day in 2004, when a massive earthquake beneath the Indian Ocean generated a tsunami that overwhelmed coastal areas as far apart as East Africa and Malaysia. Localities closest to the epicenter for the most part suffered the worst damage, as waves in excess of 15 meters smashed into the coast, in some places running for kilometers inland.

In the aftermath of the tsunami, researchers looked into the circumstances that shaped different impacts. While it is difficult to generalize, some broadly consistent findings emerged.

A review of the impacts along the coast of India found that many places that suffered lower levels of damage were protected by areas of mangrove vegetation. The same was found around the Sundarbans of Bangladesh. In Thailand, the west coast island chain of Surin escaped major damage because there was protection from a ring of coral reefs and mangroves. Similar findings emerged in places much closer to the epicenter, including Simeulue Island, where the death toll was low in part because of an emerald fringe of dense mangrove forest.

In Sri Lanka, too, those areas protected by natural shields in the form of mangroves and coral reefs fared better in the face of the tsunami than those without. Where reefs were absent, or had been degraded by dynamite fishing, the damage tended to be worse. Some local eyewitness accounts tell of how there was a visible reduction in the speed of the approaching wall of water as it came near to a coral reef. By acting as submerged breakwaters, these natural systems caused drastic wave attenuation, reducing the size of the tsunami by as much as 80 percent. Where the coral had been removed, the approaching water was uninterrupted and flowed more quickly, tending to cause more damage when it hit the land.

At Hikkaduwa on the southwest corner of Sri Lanka, the coral reefs are in better condition than many in the country, not least because they are protected in a marine park. In this area the tsunami waves that hit the land were about 2-3 meters high and caused damage to a distance of about 50 meters inland. At Peraliya by contrast, just 3 kilometers to the north, where the reefs had been widely damaged by coral mining, waves were 10 meters high and flooding occurred up to 1,500 meters inland. This beneficial effect of coral reefs might also explain why the very low-lying Maldive islands were spared from destruction, even though they were right in the path of the tsunami.

Mangroves, in common with other wetland (including those around New Orleans), also have important roles in helping coastal areas to adjust to more gradual change, including the sea level rise that is now accompanying climate change. The mechanism for this is quite simple. As tidal water passes twice a day into the tangled stems of a mangrove forest, so the waves are calmed and the flow is slowed. As this happens, the sediments carried in the water can sink. Around the mouth of massive sediment heavy rivers like the Ganges, the mangroves help to build up the coast with material carried from inland – which in this case is from as far as the high Himalayas. The muddy deposits under the trees testify to how the trees are removing the power of the water, and in the process building up the coast, including in those areas where the land is slowly sinking. The annual sedimentation rate depends on local conditions and the quality of the mangrove forest, but up to 8 millimeters per year is typical – in most cases enough to cope with both sinking land and sea level rise. Replacing this kind of natural accretion, whereby coasts evolve a new shape in response to changing conditions, with engineered structures made from rock and concrete, is often a very expensive choice, and also a less effective one.

The researchers who delved into the effects of hurricanes Katrina and Rita wrote that there was "conclusive evidence of surge reduction by wetlands as water moved inland" and said that their findings "suggest that coastal wetlands are capable of attenuating hurricane waves and protecting coastal areas against surge damage." This specific conclusion is backed by a broad estimate from the United Nations Environment Program suggesting that reefs and mangroves, depending on their health and physical characteristics, typically take out between 70 and 90 percent of the energy in wind generated waves.

In northern latitudes too, living systems in the sea protect coastal areas. On an island in Barkley Sound off the coast of Vancouver Island in British Columbia, Canada, archaeologists found evidence for an historic Native American village. They had not looked there before, because it looked too inhospitable, with big boulders on an exposed beach, it was unlike the sheltered sandy coves where the remains of such settlements are normally found. Why this was the case is thought to be down to those unique marine mammals we met earlier: sea otters.

200 years ago this coast would have looked quite different. Back then, when sea otter populations were thriving, the seaward side of the site was fronted by dense kelp forests. The seaweeds slowed the waves and dampened the ocean's power, making the beach more

sheltered and allowing sand to build up. With no otters the shore was open to the full power of the Pacific Ocean, which took away the sand to leave only the big boulders that lie there today.

These and other findings, supported with evidence and anecdote, sound like common sense, yet there is continuing loss and damage to these systems. More than a third of the world's original area of mangrove has already gone, and some countries have lost up to four-fifths of what they had in recent times. It is a pace of loss even faster than that of tropical rainforests. In the Americas the annual rate of clearance stands at about 3.6 percent. In some locations it is much worse.

When it comes to the world's coral reefs, about a third is believed to be seriously damaged already and the extinction rate for corals is thought to be increasing faster than for any other group of living organisms. Direct damage, overfishing and climate change are among the pressures that could lead to much of what remains being lost, even within the next decade.

Recent research reveals how different kinds of pollution are also taking a toll. For example, coral is affected by some of the substances in sunscreen. Each year some 4,000 to 6,000 tonnes of sunscreen is released into the sea in reef areas. The creams cause the corals to become stressed and for the symbiotic relationship that exists between the coral animals and the algae that live with them, to break down. This is known as bleaching and up to 10 percent of the world's reefs are at risk from this kind of pollution. Pesticides, hydrocarbons and other contaminants can also hasten the bleaching process.

Some coral reefs are also suffering from changes to the normal ecological relationships that enable them to thrive. To an extent, this is reminiscent of what happened in *Biosphere 2*. In that case, the ocean biome created by John Allen and his team had to be maintained by crew members who removed algae from the corals before they became smothered and died.

In natural reefs different species undertake this kind of task. Across many parts of the world a group of creatures called rabbitfish do the job. As the name suggests, these brightly colored animals, which grow to about 40 centimeters, are voracious herbivores, capable of stripping reefs of algae that might otherwise smother growing coral and kill it. Algae can be especially problematic when reefs have been damaged, for example, by a storm. Normally reefs will recover from such trauma, and the services they provide will be restored, but not if fast growing algae take hold and choke the young corals. However, rabbitfish are good to eat, which has made them a target species for fishing, to the point that the beneficial services they provide have in some places been largely removed.

While lots of fish species eat such weedy growth, researchers looking at the ecology of the Great Barrier Reef in Australia found that the arrival of rabbitfish transformed areas that had become choked with weeds. Rebecca Fox, one of the research team, said that the work done by such grazing fish is declining globally: "In Australia these herbivore fish populations are still in fairly good shape, but around the world, as the big predators are fished out, local fishermen are targeting the herbivores. In Hawaii, the Caribbean, Indonesia, Micronesia and French Polynesia, there are reports of declines in herbivore numbers of up to 90 percent ... by killing them, we may be unwittingly eliminating the very thing which enables coral reefs to bounce back from the sort of shocks to which human activity exposes them." Similarly disruptive can be the capture and removal of triggerfish, in this case leading to an explosion in urchins, which in turn eat their way through the living fabric of reefs. As was the case with Pacific Ocean sea otters, the departure of these urchin predators can lead to profound ecological shifts, and in this case, too, causing changes that have tangible economic implications. The loss of oyster reefs has also caused some coasts to become more exposed, and that is one reason why in many areas, especially in the USA, they are being restored. Some have speculated on the extent to which the impact of super storm Sandy might have been reduced when it hit New York in 2012 had the region's once extensive oyster reefs still been intact.

Eco-Insurance

There is clearly huge worth to world economies in the salt marshes, mangroves, coral reefs, oyster beds and other coastal systems. The problem is that the extreme conditions which enable us to see their value in protecting life and property are rare, so we continue to permit them to be cleared, drained, disrupted and blown apart, for short-term gain.

One approach that might alter our perception is to see these systems as a good insurance policy – providing cover when required, even if rarely (or, better still) never needed. After all, we pay substantial premiums for insurance policies year after year, often without making a claim. But what is the insurance value of these coastal ecosystems? It's hard to come up with the numbers, but several researchers have made an attempt, and have come up with staggeringly big sums.

The Caribbean coast of Central America is famed for gorgeous coral reefs and mangrove fringed coasts. The world's second largest barrier reef is found here, along the coast of Belize, and it was this region that inspired the analogues of the coral reef and mangrove systems included in *Biosphere 2*.

In 2008 a major study carried out by the World Resources Institute and WWF set out to assess the services provided by the region's natural systems to three economically important sectors: tourism, shoreline protection and fisheries. In relation to each of these, a high and low estimate was prepared. When it came to fisheries it was reckoned that the reef and mangroves were worth about $14-16 million per year, while in terms of tourism they were worth more than ten times as much again – $150-196 million – the equivalent of some 12-15 percent of the country's GDP (in 2007).

But both these were dwarfed by the value provided in the form of coastal protection. The researchers estimated that Belize's coral reefs provide service worth $120-180 million in avoided damages per year, while the mangroves that protect the coastline from both waves and storm surge added about another $111-167 million in protection value. That is a combined total range of $231-347 million per year.

Taken together (and even without looking at the additional value of conserving species and holding carbon) Belize's coral reef and mangroves were thus estimated to be worth $395-559 million per year. In 2009 the economy of Belize was estimated to be around $1.468 billion, so at the bottom end of the range these natural systems were providing services with an estimated value in excess of a quarter of the country's GDP. As noted in the last chapter, some of the world's coastal ecosystems are pretty rare, with much less coverage than, for example, the tropical and subtropical forests. Coral reefs cover just 1.2 percent of the world's continental shelf, while that of mangroves is even smaller, with a total area about equivalent to the size of England. Compared to the size of the globe, that is not much at all.

What all of these ecosystems are worth in financial terms will, of course, vary from place to place, depending on the vulnerability of the coast on which they are located, or the state of the tourism industry. That said, Robert Costanza's (deliberately cautious) analysis for *Nature* suggested that the annual value per square kilometer of coral ranges from $100,000 to $600,000 and, for mangroves between $200,000 and $900,000. These figures take account of shoreline protection and maintaining fisheries productivity, and in the case of reefs, their additional value for tourism and recreation.

The natural services provided by mangroves and corals are most obvious in poorer and smaller countries. Of the small island countries classified as a part of the developing world, some 90 percent have coral reefs, while more than three-quarters have mangroves. There is a cost to societies in maintaining these systems – for example, in establishing and policing of national parks – but case after case confirms how this outlay is dwarfed by the benefits achieved. For example, the average cost of running a marine protected area is $775 per square kilometer per year, or under 1 percent of the estimated global value of 1 square kilometer of reef or mangrove.

But how many wetlands along vulnerable coasts have been protected for this reason? The answer is, not much. Many have instead been cleared and drained, to make way for, among other things, shrimp rearing ponds, ports and tourism development. That trend continues, and in some places still accelerates. And, for the people doing the clearing and conversion, it is for perfectly rational reasons that they do it: to earn livelihoods. One place where this can be seen very clearly is in relation to the shrimp-rearing industry.

From Ecuador to Malaysia and Madagascar to Thailand, large areas of coastal mangroves have been replaced with ponds for rearing shrimps. This is in response to rising market demand, which has rocketed in recent decades. It has created lucrative businesses and jobs, earned income from exports and generated taxes. But does it really make economic sense for countries to sacrifice their thin fringe of coastal forests for such activities?

In southern Thailand, among the islands and intricate coast that was the dramatic setting for the film *The Beach*, there are still extensive areas of mangrove. Two decades ago there was a lot more of it, until large-scale removal got underway to make space for shrimp ponds. Fed on meal made in part with the "trash fish" that includes the young of large commercially important species, and with ponds treated with copious quantities of fungicides and antibiotics to control disease, the ecological wisdom of this industry has for some time been subject to critical scrutiny.

The short-term economic gains have, however, been so great as to discourage most governments from taking decisive action to even properly plan its expansion, never mind limit its growth. In Thailand, shrimp farmers can make ten times the country's average wage, while total industry revenue is nearly $1.5 billion.

While at one level an economic success story, analysis of the full costs associated with the Thai shrimp industry casts an interesting light on what the real costs and benefits might be. Research carried out for the Prince of Wale's International Sustainability Unit (on which I helped as an adviser) calculated that when all of the economic benefits and costs were taken into account, instead of providing overall economic gain, the industry was found to be generating a net economic loss to the world of $262 million annually. The costs that contributed to this estimate were made up of various kinds of ecological damage, including the loss of fish spawning areas, carbon dioxide released as the forests are cleared, water pollution and diminished coastal protection.

There is no reason to believe the economics are working much better in any of the other countries that have seen a rapid expansion of shrimp farming at the expense of mangroves. Thus the choices that favor the short-term interests of shrimp farmers have trumped longer-term national (and global) economic interests. Depending on the magnitude of sea level rise that accompanies climate change it is estimated that by the 2040s the annual cost of coastal adaptation could reach $26-89 billion, making the protection and enhancement of natural coastal features a highly practical question with major economic ramifications. The vulnerability of coasts to sea level rise will not only increase because of the loss of mangroves and corals, but also peatlands. For example, along the coast of the Malaysian state of Sarawak on the vast island of Borneo, the loss of peatlands could cause the large-scale loss of land to the sea. Peatland forests are being cleared away here to make way for palm oil plantations. As happened in the Cambridgeshire Fens, this will lead to the land shrinking, and across large areas to below present, never mind future, sea level. One estimate suggests that up to 10 percent of the area of Sarawak could be flooded. And it is not only coastal areas that are at increased flood risk due to changes in natural systems.

Floods and Forests

On June 26, 2007, the BBC evening news showed Sea King helicopters rescuing stranded people from the roofs of city center buildings engulfed by flash floods. For British viewers, the shock factor was that the pictures were not coming from some far-flung tsunami or hurricane hit region on the other side of the world, they were from the English city of Sheffield.

The River Don had burst its banks following a mass of water streaming down from the Pennine hills after torrential rains. And Sheffield was not alone. That summer saw widespread flooding across the UK. Many areas were hit by what were variously described as once-in-150-year or once-in-200-year events. Billions of pounds' worth of damage was caused.

The UK's 2007 floods followed a string of extreme rainfall events and have been followed by more — including the destructive flood that hit the northern town of Cockermouth on November 19, 2009. That day, the rivers Derwent and Cocker swept through thousands of homes and businesses, causing massive damage. Bridges and roads were swept away. In June 2012 there were again pictures of RAF helicopters rescuing British civilians from floods, this time in west Wales.

More frequent large-scale floods are a worldwide trend, highlighted by data collected by climatologists and information gathered by insurance companies. In September 2009, the Philippines was hit by severe storms, causing widespread flooding, damage to property and loss of life. In 2010 and 2011, massive floods devastated Colombia. In 2010, northeast Brazil was affected by damaging floods, followed by more the following year in the southeast of the country, causing 500 deaths. In 2010, a pulse of flood water flowed off the Himalayas and down the valley of the River Indus to cause devastation across much of Pakistan and displace some 20 million people; there were more the following year, this time causing about 8 million people to leave their homes. In autumn 2011, as I flew in and out of Bangkok en route to a work assignment in Vietnam, the airliner I was aboard passed across floodwaters that appeared as an ocean covering a vast area of southern Thailand. At the end of 2011, flash floods swept across the Philippine island of Mindanao, killing more than 1,000 people and making 30,000 more homeless.

Global reinsurance catastrophy reports confirm that 2011 was the year with the costliest disasters on record. According to insurance industry leaders, this is placing pressure on the availability and affordability of insurance, and thereby shifting greater risk exposure on to governments and individuals.

While these flooding events occurred in countries across the world, they had something in common: all of them are believed to have been made worse by the loss of natural habitats, especially forests and wetlands. These and other natural systems intercept and hold water, releasing it gradually, and in the process moderating the flow of water. And, while the world speculates as to whether the flooding events are a sign of climate changes (and collectively they do appear to mark a change in average conditions over previous decades), less time has been spent on looking at other contributory factors that are more measurable and attributable.

However, as with the impacts of Hurricane Katrina and the effects of the Asian tsunami, researchers have looked at the role of natural habitats in shaping the outcomes of extreme floods at more inland locations.

One of the largest storms of modern times – and the second deadliest Atlantic hurricane in history – was Hurricane Mitch. This hit a large swath of Central America in 1998, its effect worsened by the slow speed with which it crossed the region, as prolonged and intense rains lasted for six days. It led to the death of about 18,000 people, left nearly 3 million homeless and caused damage to housing, infrastructure and farming estimated at about $6 billion. Carlos Flores Facussé, then President of Honduras, said that the storm had destroyed fifty years of progress.

I visited neighboring El Salvador in 2001 and heard first-hand accounts of the damage caused by Mitch. I was told that the main cause of loss of life was mudslides, and that these were made worse because tree cover had been removed. The soil was washed from unprotected hillsides with the huge volume of rain turning it into great torrents of mud that then swept all before them, including thousands of homes, and sometimes with the people still inside them.

El Salvador has only about 2 percent of its original forest cover and efforts to save what is left and to restore lost forest are hampered by rural poverty and the fact that many people depend on subsistence farming and cutting timber for fuel. The deforestation makes the poverty worse, as it leads to levels of soil loss that have left much of the country unsuitable for farming.

After the storm, aerial surveys revealed how most of the land-slides had occurred on hills cleared of vegetation for agriculture and human settlements. In forested places, including where crops like coffee and cocoa were grown under the shade of canopy trees, few landslides occurred. These are common sense observations but they have been backed up by a number of scientific investigations.

One piece of work, led by Australian Professor Corey Bradshaw, looked at possible linkages between deforestation and the risk and severity of floods across the developing world. Using data collected from fifty-six countries between 1990 and 2000, his research team found that where natural forest had largely gone, flood frequency went up. Their finding is of more than passing interest. During that decade some 320 million people were displaced by floods, 100,000 were killed (nearly a fifth of them by Hurricane Mitch) and economic damages were estimated to have exceeded $1.151 trillion. Even if the loss of natural habitats accounted for only 10 percent of the property damage and loss of life, that is still a huge impact.

Bradshaw and his team concluded that "unabated loss of forests may increase or exacerbate the number of flood-related disasters, negatively impact millions of poor people, and inflict trillions of dollars in damage in disadvantaged economies over the coming decades." Over a decade has elapsed since 2000, the point at which their data ended, and it would appear that so far they were right.

This is serious enough, but it is expected that extreme conditions are set to become more common. In late 2011 the Intergovernmental Panel on Climate Change (IPCC) released a special report linking the ongoing release of greenhouse gas emissions with extreme weather events, such as hottest days and droughts, high coastal waters and cyclone activity, heat waves and heavy rain. The authors of this report estimated that the frequency of extreme weather events could increase by a factor of about four in the next thirty to forty years and by a factor of ten by the end of the century. Extreme weather events are also projected to last longer and increase in intensity. If at the same time natural systems from mangroves to coral reefs and from upland forest to peat bogs become more degraded, so the vulnerability of societies will increase.

A paper presented in the journal *Science* in April 2012 demonstrated how the global water cycle has already speeded up, seen in how more water is evaporating from the oceans compared with a few decades ago. And, with warmer air able to hold more water, flooding can be expected more often. After all, what goes up generally comes down again.

No wonder many analysts these days speak of how this combination of factors will lead to reduced resilience, in other words will diminish our ability to cope with shocks, to absorb them and to recover from them. When the ability of natural systems to ameliorate extreme weather and other shocks is reduced, then the resilience of human societies is reduced, too.

Bioshield

While much of the climate change work that I and many others have been involved with has been about cutting emissions, there is an emerging view that we must increasingly focus on finding ways to cope with what are now inevitable impacts. We are entering a period of consequences, and we will need all the protection we can get. Properly maintained and managed natural systems are often more effective (and often cheaper) than concrete engineering works.

We need urgently to harness natural systems, and use wetlands, reefs, forests and others to buffer societies from shock.

Higher average temperatures will not only make droughts and floods more frequent; more heat waves are expected, too, posing threats to public health. One way of dealing with that is to invest in more air conditioning equipment; another is to plant trees, which create a cooling effect by reflecting heat, producing shade and evaporating water. In cities, trees can make a big difference: one UK calculation suggests that a 10 percent increase in tree cover would reduce the surface temperature of Manchester and London during heat waves by 3-4°C. Trees planted around buildings can reduce the need for energy hungry air conditioning by a third.

Some countries have spotted the economically compelling logic of harnessing Nature for protection and insurance, and are investing in ecosystem restoration to prepare for more extreme conditions. For example, the governments of the Indian states of Tamil Nadu and Kerala are investing about $45 and $9 million respectively in replanting coastal mangrove forests. Other Asian governments are adopting similar strategies, including those of Malaysia, Banda Aceh in Indonesia and Thailand.

Major companies, too, are playing a part. The French multinational Danone (whom I have advised) is investing in habitat restoration through its Fund for Nature. This includes the replanting of coastal mangroves in Senegal and India. This work emerged as a result of the company's Evian brand looking for ways of creating carbon absorbing benefits to compensate for that released as a result of the packaging and transport of its product. Not only does the company contribute something important on the carbon reduction agenda, it is also helping to provide what some local Bangladeshis call a "bioshield" that will assist in protecting against the effects of extreme weather.

Replanting mangroves is not that straightforward, however. Seedlings need to be grown ready for planting, the correct species for the particular location is required and young trees need to be put in the right place. And, as well as getting the science right, experience shows that when local communities are involved, success in re-establishing mangroves is more likely.

Coral reef restoration projects are also underway, although generally on a smaller scale than the more ambitious mangrove replanting programs. From Bali and Fiji to the Bahamas and from the Red Sea and Panama to Thailand, different methods are being trialed so as to refine methods for reef rebuilding.

One promising trial called Biorock™ Process involves the creation of submerged structures whose metal frames are coated with limestone. This is achieved by passing an electrical current through the water which causes calcium carbonate to accumulate on the metal, forming a material as strong as concrete. Marine life begins to attach to this and corals can be transplanted. Reefs built with this simple technology are now growing in Maldives, Seychelles, Thailand, Indonesia, Papua New Guinea, Mexico, Panama and the Saya de Malha Banks in the Indian Ocean. The process can be made more benign with the use of renewable electricity sources, such as wind power.

Coral reef and mangrove restoration is an expensive and sometimes complicated undertaking, however. By far the best option is to ensure that they are not degraded or removed in the first place. That means, among many other things, taking action to curtail conversion of coastal wetlands to shrimp rearing ponds and to end the use of dynamite and cyanide for fishing in areas of coral reef. Fortunately there are more and more sources of research that tell of the economic sense of doing this. One study in Vietnam concluded that planting and protecting nearly 12,000 hectares of mangroves cost just over $1 million, but saved annual expenditures on maintaining sea defenses of well over $7 million.

Bogs, Woodlands and Wetlands

It is also possible to take steps to restore flood moderating natural systems on land, and this too is already underway. As we have seen, the conservation of the páramo above Bogotá (coupled with measures to cut down soil erosion lower down) will not only help to protect the supply of clean water, but also slow down the runoff of rainwater, and thus cut down on the risk of flooding. Recent flooding in Colombia is estimated to have caused damage equivalent to about 2.5 percent of GDP. That is a huge amount, and the benefit of softening the impact of extreme conditions seems set to become progressively more valuable as the climate continues to change.

In England, too, following the damaging floods of recent years, there has been a shift of emphasis. Not only is there now a focus on the maintenance and construction of hard engineered structures to prevent the flooding of vulnerable areas, but also more meaningful consideration of what might be done by working more thoughtfully with Nature.

For example, a partnership of local and national organizations are working together to curb flood risk in the flood prone Yorkshire town of Pickering through enhancing the natural ability of the surrounding landscape to hold water. Future deluges will be reduced in both frequency and scale through the planting of woods. Trees will hold back and slow the flow of water from nearby hills. There is also associated work to change patterns of drainage in upland areas. This includes the filling in of canalized watercourses and the return to more natural streams connected to small floodplains.

Further south, in the Pennine hills of the Peak District, a new partnership called Moors for the Future has set out to restore large tracts of upland blanket bog. The plan is to restore areas of bare and eroding peat – some 30 square kilometers – to habitat with growing vegetation, including the recreation of sphagnum moss bogs and woodlands. Gullies will be blocked and heavily eroded footpaths repaired.

These blanket bogs that clothe the "backbone of England" are in some ways like the páramo of the high Andes, like a huge living sponge. Not only will this reduce the risk of the kind of floods that led to Sheffield residents being rescued by helicopter; it will assist in the conservation of moorland wildlife and enhance the carbon sink that is in the peaty soils there. The moderating effect on the water cycle will also help to ensure water security in a time of climate change. And there are social benefits, too. The 16 million people living within one hour of these uplands will find areas of wild Nature in which to build physical and psychological wellbeing. Others from further afield, coming to experience extensive landscapes, will bring the economic benefits of tourism.

Lowland wetlands can also play an important part in flood alleviation. One that I know well and which will be more and more important is the Great Fen in north Cambridgeshire. The region's Wildlife Trust has embarked on an ambitious wetland restoration program here, centered on the Nature reserve of Woodwalton Fen. This place has a very special history, in being one of the first Nature reserves established in the UK. It was bought by conservation pioneer Charles Rothschild in 1912 and became one of just a few tiny areas of the ancient fen wetland to survive the onslaught of modern farming.

The Wildlife Trust aims to restore 37 square kilometers of fenland habitat and to unite two of the four last fen fragments – Woodwalton

Fen and Holme Fen (where that metal post was driven into the ground in 1852) — as a single area of wetland habitat. The enhanced floodwater storage capacity that will be created with the restored wetland will reduce the flood risk to surrounding farmland and communities. And by rewetting dried and exposed peat soils, and thereby halting their erosion, the release each year of some 350,000 tonnes of carbon dioxide will be ended. The extended wetland will also enable wildlife to adapt to climate change by providing a larger refuge.

Restoration programs like the Great Fen are sometimes received by critics as nostalgic attempts by conservationists to recreate lost landscapes. As this and other initiatives demonstrate, however, they are very much about meeting modern challenges. The large new Nature reserve that will result from the initiative will help us to meet other challenges as well, not least in promoting the health of people. The Great Fen is in one of the fastest growing parts of the UK, where the economy has remained strong and more housing is planned. In a part of the country where opportunities to experience wild places is limited, it will provide an amenity offering the chance to see rare wildlife, to walk in open wild areas and to travel through a large natural area by bike, horse and boat.

Providing opportunities for more people to enjoy time outdoors certainly has value for the people doing the walking, riding and other activities, and in recent years there has been a fast growing body of evidence to show that this has considerable economic value as well.

London's Olympic Park had Nature built into its fabric.

CHAPTER 10

NATURAL
HEALTH SERVICE

$12 Million: Health benefit of 10 percent more cycling in Copenhagen

£105 Billion: Annual costs caused by mental illness in England

£630 Million: Annual cost of maintaining 27,000 parks and green spaces in the UK

MOST WESTERN COUNTRIES actively protect their citizens from the health threatening consequences of environmental pollution, and have done so for several decades. Some of the worst toxic pesticides have been phased out, power stations cleaned up and vehicle engines regulated to cut harmful emissions. A lot of this has been done in the name of public health, and a great deal has been achieved in making air, land and water cleaner.

I have been involved with work leading in this direction much of my life. During my time at Friends of the Earth, among many other things, I helped publish pollution inventories, urged for stronger pesticide regulations and lobbied for laws for cleaner engines. The results were often positive. I was, however, always aware of far more profound implications of our relationships with Nature, and knew that we needed to look beyond protecting ourselves from being poisoned by the substances we had collectively released into it. Times are changing, though, and what was once a rather fringe discourse is now entering the mainstream.

Evidence for this can be found at Gateshead, a town by the side of the River Tyne in the northeast of England. This region was once an industrial powerhouse, a hub for coal mining, shipbuilding and steelworks. I spent a fair bit of time in that part of England, campaigning on pollution issues, not least – through making links between emissions from the many chemical works and factories there – the health of local people.

Gateshead is in one of the more socially challenged regions of the United Kingdom, and the pollution was one reason why health inequalities were worsening. Just over a third of the population live in neighborhoods classified as among the most deprived in England. Early deaths from heart disease, stroke and cancer are above the national norm, while life expectancy is below average. It sounds like the kind of place where medical facilities and drugs will be in big demand, and in this social context indeed they are. But what about woodlands? To some eyes, trees might not seem a top requirement in alleviating social deprivation, but in 2004 that is exactly what a group of public agencies decided would help. They chose a place called Chopwell Wood – 360 hectares of mixed woodland – for an experimental project, run by a partnership that included the Forestry Commission and the National Health Service.

One of the Chopwell Wood schemes enabled family doctors to send patients there to take exercise – to cycle, walk, participate in tai chi classes, or help with conservation works aimed at increasing the ecological value of the woodland. The idea was that physical activity in the open air would be effective because it would be more popular, with more patients turning up for their treatment for longer. It was hoped that the scheme would be helpful in treating patients with mild depression, high blood pressure or a weight problem. The alternative on offer was a thirteen-week period attending a leisure center or gym.

While the completion rate of patients referred for indoor exercise was low, with only about a third completing the course, over 90 percent of people referred to Chopwell Wood finished their program. And the patients said that being in the wood did them good. Some 99 percent felt that visiting the woods had a positive impact on their health and wellbeing; 60 percent thought it benefited their health through undertaking physical activity, while 40 percent said it improved both their mental and physical health. Patients emphasized the attraction of being out in woodland surroundings, relaxing and being physically active.

So what is going on here? Why were patients apparently more disposed to activity in a wood compared to a gym? Might it be because there are some inherent benefits from interactions with Nature, something that connects with us in beneficial ways because of inbuilt human affinities with the natural world? There is an increasing body of evidence to show that this is indeed the case. While scientific research demonstrating the therapeutic benefits of Nature accumulates year after year, some important findings have been around for a while.

One famous piece of work dates back to the 1980s. It was founded on data collected in a Pennsylvania hospital between 1972 and 1981 and showed how the speed of recovery after gall-bladder surgery depended in part on whether patients could see trees and greenery. Twenty-three surgical patients who had undergone the procedure were assigned to rooms with windows looking out on a natural scene. This group had shorter stays in hospital after surgery, received fewer negative comments in nurses' evaluations on how they were doing and took fewer potent painkillers, compared with twenty-three matched patients in similar rooms with windows facing a brick building wall.

Research collected in a hospital environment to demonstrate a relationship between Nature and wellbeing is one thing; finding a link between Nature and health in the normal run of things is more difficult. But researchers have done it.

Dutch scientist Jolanda Maas and a team at the Netherlands Institute for Health Services Research looked for relationships that might exist between urban green spaces and health, and if there was such a link to see how strong it was. Data was gathered from more than a quarter of a million people using a questionnaire. Those taking part were asked about their general health and how they felt. The data was analyzed in comparison to the distance they lived from green spaces, including urban parks, natural areas and farmland.

From this, some striking findings emerged. If green space was nearby, then people rated their own health as better than those who didn't have such spaces near their home. For example, in areas where 90 percent of the local space was green, 10.2 percent of the residents felt unhealthy. In areas where 10 percent of the environment was green, then 15.5 percent of the residents said they felt unhealthy. This kind of relationship was consistent across different neighborhoods, but green space was found to be especially beneficial for people on lower incomes.

The difference between 10 and 15 percent, when scaled up to the populations of cities (or countries), can mean millions of people feeling less healthy than they otherwise might, and the economic consequences of that are considerable, both in terms of the strain on health services and lost days at work. For example, and according to the Mental Health Foundation, in England the cost of mental health problems is estimated at around £105 billion per year.

Although planners sometimes struggle to place values on green spaces, it is interesting to see that markets clearly do so. For example, a 2003 survey in London found that a 1 percent difference in the amount of green space in a locality was associated with 0.3-0.5 percent increase in the average house price. The authors of the Dutch study concluded that "green space seems to be more than just a luxury ... and should be allocated a more central position in spatial planning policy."

While Maas and her colleagues looked at the distance of all kinds of green spaces from people's homes, a British research project set out to see if a greater variety of wildlife in our surroundings has psychological benefits. A team led by Richard Fuller at the University of Sheffield looked at local differences in the city. In a segment going from the center to the western suburbs, fifteen green spaces were surveyed for their wildlife diversity. Researchers measured the richness of the plant species; butterflies were identified and recorded, and so were the birds. In order to understand the reaction of people to different levels of wildlife diversity, interviews were conducted with over 300 users, seeking information on how the green spaces were regarded as important for psychological restoration, positive emotional bonds and their sense of identity.

The study found a measureable positive link between the level of species diversity and the psychological benefit expressed by those interviewed. The number of habitats found in a particular place was correlated with reflection and personal identity, plant variety tended to be associated with an ability to reflect, and birds were associated with emotional attachment. The researchers suggested that urban green spaces might best be managed to create mosaics of habitat so as to enhance wildlife variety, and thus maximize the psychological benefits for users.

There is considerable evidence, too, showing how exposure to Nature is associated with reduced stress and improved work performance. More specific studies have also demonstrated a link with lower blood pressure among dental patients, fewer reports of ill health among prisoners, increased self-discipline among inner city girls and reduced mortality among elderly people. Children with attention deficit disorder have been found to show significant improvements if they play in natural areas, or even have views of trees and grass outside their home.

One study found that for every 10 percent increase in green space, communities can show a reduction in health complaints equivalent to a reduction in average age of five years. Another found that exercise in a green environment can create an immediate improvement in self-esteem. Several researchers have demonstrated how office workers experience lower job stress, higher job satisfaction and fewer illnesses if they have views of natural areas than if they did not. Another study measured driver stress, as expressed in blood pressure, heart rate and the state of their nervous system, and found that these indicators went down when roads passed through green areas and when there were trees by the highway.

Children suffering from stressful situations recover more quickly when they are in areas with access to Nature. As is the case with other studies, the benefit is seen most clearly among low income and socially deprived groups.

Reviewers looking at the accumulated literature have concluded that being exposed to Nature reduces anger and anxiety, sustains attention and interest, and enhances feelings of pleasure. Scientists have shown how the psychological response to Nature can involve feelings of pleasure, "relaxed wakefulness," and reductions in negative emotions, such as anxiety and anger. Pioneers in this area of research are Rachel and Stephen Kaplan of the University of Michigan. They have described the kinds of "restorative environments" that foster recovery from mental fatigue and their correlation with more natural settings.

On top of the gains that individuals can get from Nature, there is evidence pointing to social benefits. For example, community cohesion has been found to increase where there are more trees. In one study on a Chicago housing estate people knew more of their neighbors when there were more trees where they lived; in areas dominated by concrete, people interacted less. There was also less domestic violence in areas with more trees.

Dr. William Bird, a British family doctor based in Oxfordshire, became convinced of the therapeutic value of Nature while running a diabetic clinic in the early 1990s. He saw how green space around where his patients lived was not being used by them, and as a result he started a program of health walks in natural areas. This was a great success and the idea has since grown to become a national scheme, run by the British Heart Foundation, a major charity.

Initially, Bird found it hard to get the medical mainstream to take the idea seriously. However, when he began working at Natural England, an official agency, he was able to forge links with the Department of Health, and a route into policymaking. I met with him in the members lounge at the British Medical Association, where he explained to me the linkages he'd found between improved health and access to Nature.

The most significant health benefit of Nature, according to Bird, is a reduction in stress. "There is evidence that most of the chronic diseases are linked to stress – dementia, diabetes, cardiovascular problems, depression and so on," he told me. "And it seems to me that stress has been totally underestimated. Chronic daily stress is the problem. Stress out of control."

Bird explained how public health interventions based on more access and exposure to Nature could be a cheap way of taking pressure off health budgets. He showed me the results of various studies that demonstrate how the benefits are especially marked among lower-income groups: "Stress comes from uncertainty and fear and all those things are more pronounced in lower income groups. There is research to show that, even if you don't smoke or drink and take proper exercise every day, there is a still a massive gap in health outcomes between the income groups, and I think this is down to stress and people living in fear. Their environment is rubbish, there is no social cohesion and they have no job, and so the stress goes up."

"The stress has a direct effect at the cellular level. It also affects behavior. What do people do when they are stressed? We eat too much, smoke and drink. We do all of that because of the stress. So if you live in a deprived area, then you will do all these unhealthy things as a way of treating the stress. So when people say you mustn't smoke, drink or eat too much the reaction is often to dismiss that advice, not because they disagree with it but because there is an underlying stress that is more powerful than the good advice. A connection with the natural environment helps us feel our purpose and in the process helps to reduce stress."

He went on to explain the significance of Nature to the normal development of young people, and how children living in urban areas with no access to green areas can suffer serious damage. "By denying a child access to green areas they are being denied an essential part of their physical and mental development. There are some children now who can't balance on one leg simply because they have never been on uneven ground."

One piece of research Bird carried out was to map what he called the "roaming zones" of four successive generations of a single family. This project was carried out in Sheffield and found how, over time, children's roaming range had shrunk, from the great-grandfather, to grandfather, to mother, to the present-day child.

"The first one was six miles, walking in the hills, going fishing and so on. He could roam everywhere. Then in the next generation it was down to a mile, but still with access to woods and so on. Then the mother was down to half a mile, and with Tim, the 8-year-old, it is 300 meters – which is the average of most children now. These ever decreasing circles, of course, limit access to Nature, especially if children don't have such environments in their very local area. Bird has no doubt that this is a significant social issue. "There is research from Germany and the United States which shows how children who didn't have experiences of free roaming in Nature by the age of 14 would never get it for the rest of their lives."

Bird soon found himself invited to conferences to share his insights. "There was a massive need all around the Western world for the health people to meet the Nature people." But while interest has been intense, he says there remains a mind-set issue among the medical establishment. "Some doctors feel that contact with Nature is somehow going backwards. They think that progress is a new scanner. Things are changing for the better, though. There is more acceptance that not everything can be done with technology. Obesity, depression, you can't deal with those through cutting someone open."

Economics plays a significant part in this thinking, too. Bird gave me the example of the cost of achieving a "quality adjusted life year" for patients suffering from risks of cardiovascular disease. Drugs in the form of statins cost about £9,500 per year, while exercise based activity costs about £440 – twenty times less. Moreover, statins don't work as well as walking, and cause side effects.

When juxtaposed against the hundreds of billions of dollars spent globally each year on healthcare, these and related findings have important implications, not least in how we design the places where we live, work and play. Finding an optimum accommodation between people and Nature is clearly about far more than simply cutting pollution. But why is it that we humans seem to benefit from contact with the natural world? We are, after all, now mainly urban creatures, so why does Nature still matter?

Hunter-Gatherers and Nature Deficit Disorder

Ancestors who were modern human's anatomical equivalent first appeared about 200,000 years ago. What makes us human was forged during that period, and indeed in the several millions of years of hominid evolution that took place before then, never mind the vast period of primate evolution that led to the hominid apes. Urbanization really got going only about 200 years ago, and even then only for a small minority of people (it was as recently as 2007 that, for the first time in human history, the majority of people lived in towns and cities).

The long period of human history that led us to where we are now places the recent process of urbanization into a rather different context than the decadal time scales we usually associate with the expansion of built-up areas. Even if one takes 200 years as the period in which humans have been urban (which is really an overstatement), then about 99.9 percent of human history was not urban.

Before we lived in built-up areas, we spent our time much closer to Nature – in fact, we largely lived in Nature. For most of human history our basic economic model was hunter-gathering. If for the sake of making a point we assume that this was our dominant means of meeting our needs until about 10,000 years ago, then about 95 percent of human history is accounted for by this way of living.

To be effective hunter-gatherers, people must be competent naturalists, acutely tuned to the seasons, aware of the migratory patterns of important animals and knowledgeable about the fruiting patterns of different plants. They must know about soils, spring lines and, of course, predators. Hunter-gatherer groups living in the Amazon rainforests today do not simply pass through the landscape taking what they need; they manage the forest so as to produce more of what they want, and this means planting and tending those plants which they regard as more valuable, including the trees which produce the fruit which feed the wild pigs they hunt. What looks like a natural rainforest is in part a giant, elaborate and semi-natural garden.

And when humans made the great leap from hunter-gathering to agriculture, there was still a closeness and known dependence on Nature, and this in turn affected almost every daily experience. Was the soil sufficiently fertile to produce a crop this year? Would it rain enough? Would a major pest attack cut the yield? If any of these things was misjudged, then starvation and conflict could easily be the result.

The research I have briefly summarized in this chapter provides a little insight as to the practical consequences of this. And bear in mind that the brains which do our thinking, reasoning and decision making, which bring emotion and awareness, are evolved from the brains of primates that were as much a part of the forests and plains as the grass and trees. Like it or not, we are still embedded in Nature, and it seems we still "know" that, even if in our concrete, glass and steel cities this may not be an obvious or front-of-mind fact.

William Bird talks about the small proportion of time that humans have spent living in their modern city environments and of how we have been shaped in quite profound ways by our birth and long existence in Nature. "It's like having a computer system that took about a year to develop and then somebody says in the last minute, 'Can you please change the entire thing?' There is no way."

Our inherited predisposition to associate with Nature, our basic programming, has many manifestations. For one thing, there is the great enjoyment many of us get from tending plants, whether that be in gardening, producing food in a vegetable patch or even growing house plants. Considering the time and money that many people plough into their gardens and allotments, it is clear that some very important needs are being met in the process. Are we urban gardeners compelled to grow and tend plants in part because of the momentum of our past?

The same thing might be asked in relation to many of the relationships we maintain with animals. In Western countries billions of dollars are spent on the purchase and care of pets, and with that kind of outlay some benefits are evidently being derived from the companion animals. Some of these are increasingly well understood. Cats and dogs have been shown to contribute to reduced blood pressure, increased ability to cope with stress and as therapeutic in helping with minor ailments. Our propensity to engage with animals is also seen in visiting zoos and aquariums; in the USA and Canada, more people visit such attractions than attend major sporting events.

In the UK the largest participant sport is angling. A lot of people who go fishing are not too bothered if they don't catch very much – the real attraction is being outside in Nature, interacting with it in an engaged manner. Could keeping dogs and going fishing also be echoes of our hunter-gatherer past?

Even if they are, we should not romanticize history. Most of the human story was brutal and filled with hazard, from disease and starvation. Life was short and painful and the rise of agriculture and cities is in part a means of escape – a search for comfort, convenience and longevity. And it has worked.

A review by Michael Gurven and Hillard Kaplan found that among traditional hunter-gatherer groups surviving today the average life expectancy at birth varies from 21 to 37 years. The proportion surviving to age 45 varies between 26 percent and 43 percent. If we assume that this pattern is broadly indicative of the situation over much of human history, then recent changes are a revolution for the better.

Over the last 160 years or so, the average life expectancy for a person has increased linearly at almost three months per year. Improvements in sanitation, medical science, access to drugs, nutrition and public health account for much of this change. Whereas most of us once died from infectious disease, hunger and injury, today (and in part because of our longer lives) we suffer more from chronic illnesses, including heart disease and different kinds of cancer, in turn linked to trends such as more obesity. And because of the way we live, there is also a burden of psychological and behavioral challenges. About a tenth of health disorders reported in the West today are psychological, and this is expected to rise to about 15 percent by 2020.

Such modern health challenges are not easy to address with technology alone. To cost-effectively combat heart disease, cancer and mental disorders, different approaches are needed. And the more research that is conducted, the more clearly it seems that Nature could be a large part of the solution. Yet at no point in our history have so many humans spent so little time in physical contact with animals, plants and the processes that govern the natural world. We are suffering from Nature Deficit Disorder.

Olympic Gold, Green Walls and Seoul's River

Based on the kinds of findings mentioned in this chapter, many experts conclude that encouraging time spent close to the natural world can be an effective public policy goal in promoting better health, and in particular the prevention of mental illness. In order to do this, more effective collaboration is needed between (among others) those working in the primary healthcare sector, architects, urban planners and environmental managers. Some make their point by talking about the potential to create a Natural Health Service. Treatments provided could take many forms, from tending pot plants to forests, and everything in between, including gardens, allotments and urban parks. The point is proximity and contact with Nature's different elements: animals, plants, growth, water, recycling, decomposition and the other processes which sustain the whole intricate tapestry, which in turn sustains us, physically, psychologically and (in many cases) spiritually.

Certainly my own personal experiences of exposure to Nature are very positive. Perhaps because the connections made are at a biological and even spiritual level, the experience is difficult to articulate. For me, though, walking in areas that are wild or semi-wild, looking at birds, plants, fish and animals brings a sense of optimism and wellbeing. The physical exercise is welcome, but the feeling that comes after time outside in the wild has so much more than that.

The upbeat feelings are especially enhanced if I have my dog. She walks nearby, looks back from time to time, seeking signs as to where we are going and what she should do. 30,000 years of canine domestication has not diminished her affinity with Nature, either. Sniffing, listening and looking, she is utterly tuned in. In some ways we take walks in prehistory.

Research suggests that my experience is not atypical. And as our population marches toward an expected 9-billion-plus by mid-century (with perhaps three-quarters of those people likely living in towns and cities), parks and natural areas of different kinds might well be one of our most valuable public health resources. The challenge will be making access to Nature an every-day opportunity.

William Bird is among a growing number of experts who say that part of the opportunity is in how we plan development. He says that those UK health authorities that are seeing Nature as part of the health equation are seeing results, especially when they are able to provide Nature experiences to those people who need it most. People on the lowest incomes often live in areas with no green spaces: "To fill that gap you need to work with the community to create safe Nature spaces. To ask the people who will use it how best to design it for their needs. If this is done well, it will help to increase physical activity and their behavior changes."

While the connections between Nature and wellbeing have become better documented, there are relatively few studies that quantify the economic benefits of spending time in natural areas. One city that I have come to know from many visits is Copenhagen. The Danish capital is famous for many things, including a high level of cycle use. More than a third of the people who commute into the city each day, and well over half of those who live there, cycle to work or school. They do so along nearly 400 kilometers of cycle lanes and paths. Many of these are through tree-lined routes, parks and other green areas.

It is a complex job to put specific values on this level of cycle use, but Niels Jensen, a planner working for the City of Copenhagen, has presented data estimating that a 10 percent increase in cycling in his city would save about $12 million in healthcare costs, increase productivity by about $31 million, reduce absence from work by over 3 percent, lead to an extra 61,000 years of life and 46,000 fewer years of serious illness.

One welcome success in incorporating green spaces into urban life was the 2012 Olympic Park in London. Sports facilities don't have a good record in incorporating Nature, but the London Olympics consciously set out to build Nature into the overall design. I visited a few months before the Games and wandered beside a recently established reed bed, where a small black, red and white bird flitted across a path. It was a stonechat, and the first of its kind recorded in that part of East London. Nearby, at the edge of a newly created pond, was a grey wagtail. It seemed an incongruously graceful bird to be found in a building site. But these and other creatures were not there by accident. The creation of natural habitats for local wildlife was an integral part of the overall design of the Park, with sustainable drainage incorporated with the creation of wetlands. These quickly attracted other birds, including little grebes and reed warblers, and as these areas mature other wildlife will join them.

The Olympic Village was also fringed with wetlands and wet woodlands, topped up with water from the root tops. The water flows through reeds and into a storage pond, and from there is pumped back up to water the gardens as needed. The wetlands are a magnet for wildlife and sand martin and kingfisher nesting sites have been put in to encourage some of them to stick around and breed. Some of the green areas between the venues have been seeded to create flower rich meadows.

Another development that has an understanding of the therapeutic values of Nature at the forefront of its thinking is a housing scheme on the edge of Cambridge by Swedish construction firm Skanska (whom I have advised from time to time). In addition to meeting very high standards for energy and water efficiency, the 128 dwellings here all have sight of green space from all of the windows. Trees, water and open green areas have been designed into the development, as have cycle lanes. There are pocket parks, allotments to grow food and public artworks. It is a mixed development containing substantial family homes and apartments. It is not built quite to the exacting standards that were applied to the living quarters inside *Biosphere 2,* but it is heading in the direction of achieving high construction standards and aesthetic appeal with low environmental impact, and is way ahead of most comparable sized housing schemes.

Even in shopping malls – those citadels of consumerism generally divorced from Nature – there are possibilities for positive interventions. One example is the Westfield Shopping Center at Shepherds Bush in West London. This major new retail development had to be separated from existing residential properties nearby with the construction of a long wall, but rather than create a bare concrete barrier, vulnerable to graffiti, one of the development's designers came up with the idea of a green living wall. As the name suggests, this is basically a hard structure covered with greenery on the outside.

The inspiration for the cooler north side of the wall came from the designer's home in Devon – a wet and temperate part of the British Isles where a great diversity of plants abound on shady banks. For the Westfield green wall, species of fern and snowdrops were selected. On the south side, which would receive more sun, they chose drought-resistant plants such as sedums and fescues. At 170 meters long and 4 meters high, the wall is a serious feature and a total of 5,000 modules holding 200,000 plants were needed to cover it. The wall has proved popular with shoppers, and it has been notable that restaurant premises with a view of it have been easier to let than those without – a clear reflection of its value.

Eco-Cities

On a much bigger scale, inspiration as to what can be done, not just with new developments but in reshaping whole urban environments, can be drawn from Seoul, where, in 2003, after decades of intense urbanization, the city's mayor decided to rebalance the urban landscape by restoring a stream.

The stream in question was the Cheonggyecheon River, which once divided the northern half of the city from the south, fed by tributaries flowing down from surrounding mountains. As the city expanded during the 1950s, it became an open sewer, choked with rubbish, and it was decided to bury it under a program of building works. By the early 1970s, the river route had become an elevated four-lane highway. The road helped to stimulate development and the area became one of the most congested, polluted and noisy parts of the city.

During the mayoral election campaign in 2001, candidate Lee Myung-bak presented a policy to get rid of the highway and restore the river. He said that displaced traffic could be moved on to a bus rapid transit system. He won the election and got straight to work. By 2005, the stream was back.

A 3-mile stretch of concrete and tarmac was removed and transformed into 3 miles of running freshwater fringed with pedestrian pathways, trees and green fringes. A number of fish, bird and insect species soon moved in. The stream and trees helped cool nearby areas to the point where the temperature by the river was more than 3°C lower than the city average. The great concrete edifice of the highway overpass, once a symbol of development, was thus replaced with a new emblem – a city center green space with running water.

Of course, there were voices raised against such an unusual and ambitious plan, including from some local businesses, and taking forward such a high profile and controversial proposal is full of risk for politicians.

It might, however, be a comfort for politicians wishing to promote contact with Nature that doing the right thing can be rewarded. In 2007, Lee Myung-bak was elected as President of South Korea.

Challenges on an even larger scale are presented by China, where it is estimated that 1 billion square meters of urban space are being developed each year (about half the global figure). But can China use its rapid economic development to set a different kind of example? Some people believe so, including global engineering consultancy Arup, which is presently at work on the development of Wanzhuang "eco city," near Beijing.

The idea behind this development is to place the future inhabitants of this substantial new urban area close to Nature, while retaining the benefits of urban living. To this end, Arup is proposing low tech green design employing natural lighting and ventilation and conservation of the local vegetation and wildlife. One particular problem faced by the project is that the area suffers from water shortages, which can be expected to get worse as the climate warms. To make the most of the water that is available, Arup proposed an urban agriculture and food system to produce fruit and vegetables. Peter Head, Arup designer and co-author of the plan, said that "in doing so, the strategy delivers 100 percent food security for fresh fruit and vegetables for the new community and significantly reduces water consumption, as well as doubling farming income and increasing the number of jobs related to agriculture in the area by 50 percent."

That is quite an ambition for a new urban area and, if it gets built as suggested by Head and his team, it will provide a fascinating practical example of biosphere friendly urban design. If there is anything out there right now which comes close to the ambition of *Biosphere 2,* and John Allen's aim to marry the needs of the ethnosphere with that of the biosphere, then perhaps the Wanzhuang eco-city vision is it.

Reconnection

As all these examples show a reconnection with Nature can have huge health benefits for urban populations – particularly for the least-well-off communities – and they tend to pay their way, financially, creating environments that attract investment, populations and consumers.

There is, however, another very important reason as to why societies might wish to encourage greater contact between people and Nature. You probably noticed from the previous chapters that much of what Nature does for us is in decline. If we are to halt and reverse these trends, then it will be vital to garner public support for changes in how we live.

Through many years of campaigning, I learned the hard way how science and statistics take arguments only so far, and that in order to gain support for pro-Nature policies, they need to mean something on an emotional or personal level. Increased urban-based consumerism has led to less contact with Nature, and this means relatively few people connect with their own experience about ecological matters. This change has taken place over little more than half a century.

By way of example, I recently picked up the *Observer's Book of Freshwater Fishes* – a beautifully produced little pocket guide to the eighty-two species recognized as living in the streams, rivers and lakes of the British Isles. It was printed in 1941, in the darkest days of the Second World War, and gained partial exemption from resource rationing restrictions because it was regarded as an educational book. That fact alone says something about the priorities of the day, and the extent to which knowing about Nature was seen as important, even at a time of dire national peril. What also says something is a remark made in the preface, where the author mentions some of the fish species are "familiar by sight to almost everyone." I doubt that could be written today.

The progressive distancing from Nature that has taken place in recent decades is one reason why so many people switch off when they hear about environmental issues. And when they do so, the chances of getting the changes needed to maintain what Nature does for us goes down. This is another reason why we need to invest more effort in getting people into contact with it – to literally connect with the Earth.

Getting connected is not as ambitious or hard to do as it might sound. Schools can do a great deal, as can town planners, developers and public bodies with land. It could also be a matter pursued in national policymaking. After all, it is not as if countries are strangers to using the law to make where we live healthier. In Britain in 1851, the Public Health Act was passed by Parliament to ensure water was clean. Waterborne infectious diseases were quickly reduced. In 1956, a Clean Air law was enacted, to reduce the emissions causing widespread respiratory sickness and hundreds of deaths. That worked in promoting better public health as well. Both these steps were based on a growing understanding that how we treat the environment has consequences for people's wellbeing.

Perhaps the emerging science on how contact with Nature benefits people could logically lead to equally important new laws, for example, in setting a minimum distance that homes should be from high quality green space, and new standards for urban designs to better incorporate Nature? Perhaps there should be official guidance to schools to require that children spend at least a couple of days a month undertaking activities in natural areas.

All of which may help create the political will and public consciousness necessary to put Nature center stage. However, to ensure that we maintain Nature's capacity to provide its essential services, we need to look above all to economics to achieve large-scale change.

Natural capital sustains financial capital – a rain forest in Guyana.

CHAPTER 11
FALSE ECONOMY?

$6.6 Trillion: Annual global environmental damage caused by human activities (11 percent of world GDP)

$72 Billion: Annual amount needed to avert mass extinction of animals and plants (0.12 percent of world GDP)

More Than Doubled: Forest cover and per capita GDP in Costa Rica since late 1980s

1975 WAS THE YEAR that Vietcong forces entered Saigon and the Vietnam War came to an end; the cult movie Rocky Horror Picture Show was released; and for the first time Bill Gates used the word "Microsoft." But perhaps the most significant landmark was that it seems this was the year when rising human demands on the Earth's natural systems exceeded what Nature could indefinitely supply. Until then, and through the great procession of history that took humans from upright apes to the space age, the Earth had been more or less able to support our activities and to renew itself. No longer.

There is of course a great deal of complexity behind this bald conclusion, with different trends (for example, the decline of fish stocks and build-up of carbon dioxide in the atmosphere) proceeding on different time scales. It is nonetheless helpful to have a yardstick against which to judge the extent of our overall demands on the biosphere. And things have gotten worse since. Similar analyses conclude that today, each year, we are using about one and a half times the Earth's renewable ecological capacity.

The ecosystems that naturally renew themselves, and which supply us with the huge range of economically valuable services and benefits, are sometimes seen as analogous to financial wealth, and are increasingly referred to as "natural capital." That natural capital, like financial capital, can yield dividends – in the case of Nature, in the form of the benefits and services described in earlier chapters, seen in, for example, fertile soil, clean flowing rivers, fish catches, disease control and carbon capture. But the kind of prudent behavior that spends only dividends while saving capital presently eludes us, and in many cases we have instead embarked on a short-term spree, whereby the capital itself is being blown.

And the results of this kind of mind-set are of course graphically evident in the ongoing economic crisis, caused in part through countries, companies and people consuming capital rather than living from the dividends. When the supply of capital finally became restricted, chaos ensued. Bailouts of countries and banks might have helped to manage the worst effects of a so-called credit crunch, but in the face of an emerging "Nature crunch" we have no comparable mechanism. The uncomfortable fact is that Nature does not do bailouts – at least, not on time scales that would make much difference to us.

At around 1975, when we crossed the threshold of living from dividends to eating natural capital, the world population was heading toward 4 billion people. By the mid-1980s it reached 5 billion and then, in 1999, passed the 6 billion mark. In October 2011 it reached 7 billion and is expected to be heading past 9 billion by the 2050s. Much more important than population growth, however, is the impact of continuing economic growth and the more affluent lifestyles that come with it. If we continue as now, with our quest to extend the high consumption lifestyles presently enjoyed by about 1.5 billion people to 4, 5 or 6 billion, never mind to over 9, and if we continue to do this with the means we use now, then between three and five planets' worth of capacity will be needed by the time we get to the 2050s.

It is easy to argue about statistics, but one is blindingly clear: there is only one Earth. Like *Biosphere 2,* it is a closed system and the limits to what it can sustainably provide have already been exceeded. Indeed, as human demands escalate, the Earth is in eco-logical terms shrinking, as systems on the land and in the sea are disrupted, depleted and degraded.

So with an increasingly clear body of scientific information, why are we finding it so hard to respond to such obvious danger? There are, I believe, a number of interrelated causes, some of them quite fundamentally linked to what it means to be human.

The basic settings that equipped humans with the ability to sur-vive in hunter-gatherer communities are still there, embedded deep in minds that evolved in the testing conditions of the Pleistocene – and those settings tend to favor short-term priorities and behavior. We were adapted to avoid danger and to exploit opportunities in the here and now, and our brains remain with the same basic wiring today. Our defaults require that we protect family members, achieve comfort and some level of security through juggling choices and options, and mostly in a short-term context. And we are social ani-mals, wired to value status and all that comes with it.

These settings, which determine much of our behavior, are as strong now as they ever were, and for individuals remain entirely rational. They may not, however, be so rational from the point of view of societies or in sustaining the biosphere.

For decisions that maximize individual benefit sometimes cause wider negative consequences – for the community, society or indeed much of humanity – and some of those consequences are made more abstract through being in the future.

The effects of the instinctive short-termism that dwells in our Pleistocene minds are visible in some of the trends described in preceding chapters. Overfishing, soil damage, taking too much fresh water, the clearance of forests, the hunting of rare animals and the release of pollution are generally justified by the individuals involved, because even though the economic costs to society may be high, the benefits in the here and now are judged as advantageous. And that tendency to place a premium on the short-term is highly visible in how we conduct much of our economic activity.

Capital Punishment

Modern economies are governed chiefly by two sets of decision makers: governments and private sector companies. Both are driven by short-term incentives and pressures.

Pavan Sukhdev, an Indian economist and former Head of Treasury at Deutsche Bank in India, has a long-standing passion for environmental economics, and was the leader of a high profile international process called The Economics of Ecosystems and Biodiversity (TEEB). He told me how, even though some of the corporations now understand major changes as to how they operate are necessary, "the short-term pressures are so powerful that they continue as before."

Indeed, the imperatives to achieve profit have been exacerbated in recent years, not least by the impact of technology on share trading. During the 1980s shares were generally held for years before being sold again. Today, it is more often months or even weeks or days. This is a result of more sophisticated stock management strategies, in turn made possible by lower transaction costs and information technology. Short-termism has been reinforced by the incentivization of management based on share performance and by laws that require firms to publish financial statements each quarter. These were initially intended to reduce fraud but have as well created focal points for shareholders to place even more short-term pressure on managers to increase profits (and thereby returns to shareholders), on a three-month basis. With the trends affecting Nature most meaningfully measured in decades or even centuries, this quarterly performance horizon helps put economies right outside the context of the natural systems that sustain them. No wonder, then, that, while the volume of stock trading has gone up, and the process has become ever more frantic, the attention paid to how the profits are made is generally lacking, or at best piecemeal.

Richard Burrett is another former banker with passions for natural as well as financial capital. He has a particularly clear view of where Nature meets money, through his role as co-chair of the United Nations Environment Program's (UNEP) Finance Initiative. The goal of his organization is to help the financial sector behave in more sustainable ways. To that end, he works with about 200 institutions – banks, insurance companies and different kinds of investment organizations.

Burrett emphasizes the scale of the sector and how it is fundamental to the way the world works: "Institutional investment is worth about $80 trillion. That's the money managed by pension funds, mutual funds, insurance funds, sovereign wealth funds, hedge funds and private equity. They are using shares, bonds and other financial vehicles to get a return, to make a profit." He says that "companies are beginning to think about natural capital as an underpinning service for the economy, but it's not in the mainstream." He believes part of the problem comes down to a quite fundamental misunderstanding: "In economics your stocks would be your capital and your dividends are what you would take as your income. That thinking isn't applied when it comes to Nature and instead it is seen as a set of flows rather than as capital."

In other words, natural capital in the form of forests, soils, fisheries and all the rest is being liquidated to make profits, and the process is being treated as a stream of dividends rather than the spending of capital – which is what is really going on. And the most successful businesses get at that natural capital and liquidate it for free, hastening its depletion still more rapidly. It makes great economic sense at one level, and drives a whole lot of what's going on.

The mind-set of some of those engaged in this process is from time to time glimpsed in their public pronouncements. Doan Nguyen Duc is the founder of a Vietnam based conglomerate called Hoang Anh Gia Lai. His company has grown fast and in 2012 was worth about $1 billion. Duc runs the kind of profitable enterprise that investors generally like to buy shares in. But if one looks a little closer at his approach and attitude, then surely we should have cause to question what societies recognize as good businesses. When talking about his interests in relation to land and forests Duc said in late 2009: "natural resources are limited, and I need to take them before they're gone." He is unfortunately not alone.

The approach taken towards natural capital is rather like a planetary Ponzi scheme. Ponzi schemes are fraudulent financial structures, made famous in 2009, when Bernard Madoff was convicted of running one on a multibillion dollar scale. Basically what he did was to pay interest out of capital and to continue doing that by fooling new investors to put more capital into his investment scheme. As time went on, more and more capital was needed to pay "returns" to investors, so he had to attract more and more money to keep it going. The scheme was, of course, doomed to collapse. The only question was when.

The parallels seen in stocks of natural capital being used up at the same time as being treated as some kind of dividend or interest is unfortunately apt. Bernie Madoff was sentenced to 150 years in jail, and I am sure that some investors, having lost their life savings, felt capital punishment was appropriate. But when it comes to Nature, instead of putting people in prison, the more extreme the ecological version of the Ponzi scheme becomes, the more we celebrate. The people who make it happen are awarded knighthoods and massive bonuses.

In an attempt to challenge the kind of thinking that continues to see the liquidation of natural capital as a good economic outcome, Burrett's UNEP Finance Initiative asked a research company called Trucost to look at the impacts of human activities on natural capital. Some remarkable findings came back. The top-line estimate was that global environmental damage caused by different human activities in 2008 had a financial value of about $6.6 trillion – equivalent to 11 percent of world GDP. About a third ($2.15 trillion) of this damage was caused by the world's top 3,000 companies.

Burrett explained that "One way of looking at it is to see more than 2 trillion dollars' worth of damage being caused by the largest companies as a kind of subsidy, because if these costs were fully reflected in their accounts then a lot of them wouldn't be profitable." He believes some of the "universal owners" – companies such as pension funds and insurance companies which invest right across share markets – should be very concerned about this: "There will come a point where the whole system suffers, meaning that there won't so much be winners and losers, but more like survivors and losers, because the value of the whole system will be impacted."

Burrett believes it to be quite logical that this process of using natural capital to create false profits should elicit some response from the financial sector. "It seems to me that these companies should become much more active owners and look for system-level changes, including better reporting. We need to account for everything, not just the short-term financial side, but the natural capital, too. This raises lots of challenges, of course, in putting prices on things that have historically never been priced. And it requires people in the system to understand that natural capital is the bedrock that underpins industrial, manufactured, social and financial capital."

If the Trucost study is to be believed, then Burrett's point is well made, because if we continue with business as usual then that particular piece of work estimated the annual value of damage being caused to natural systems could reach $28.6 trillion by 2050. By then the decline in natural capital will be causing inevitable feedbacks that will impact on financial capital – for example, as agricultural productivity is hit, or as water scarcity causes higher costs for industry.

At this point it might be helpful to set such multitrillion dollar figures into context. I have mentioned already the ground-breaking study by Robert Costanza and his colleagues (published in *Nature* in 1997) that set out to estimate the overall economic value of natural systems and services. What I have not yet pointed out is how this work estimated that the total annual combined economic value of Nature was nearly double global GDP. In other words, the benefits being provided by the forests, soils, wetlands, grasslands, coral reefs, mangroves, oceans and the rest are worth about twice that which is measured each year through official national accounts. As these natural assets are progressively degraded, by among other things habitat loss, overexploitation, pollution and climate change, then of course the prospects for continued growth in GDP are cut as well, as the planet scale Ponzi liquidates stock in what will become ever more self-defeating efforts to sustain "growth."

Joshua Bishop is a natural resources and environmental economist who works with WWF. He previously served as the Business and Enterprise Coordinator alongside Pavan Sukhdev on the TEEB initiative. He believes companies need to look at how they can generate net positive impacts on natural capital. He explained to me what this means: "Companies take it for granted that they have to make a positive contribution to financial capital; if you don't do that you are out of work and the CEO gets fired. Investors have a benchmark on how much you are expected to earn on the capital employed. No one thinks about Nature that way. Many talk about minimizing impacts, but rarely do they talk about making a positive contribution so that Nature is better off after an operation compared to how they found it to start with. It's no longer good enough to be reducing damage. We have reached the point whereby companies now need to make a positive impact."

Bishop's conclusions are increasingly widely shared, including among a small but important group of company chief executives. But, of course, the ability of private enterprise to move in this direction is in large part determined by the policies, laws and incentives put in place by governments. And here too there has been a tendency to see the worlds of ecology and economy as separate.

Governments undertake several vital economic functions. They raise taxes and spend money, regulate what companies can and can't do, and create the economic structures, strategies and measures that guide their agenda. But all of this is for the most part designed without much regard for what Nature does for us, and it is also driven by short-term pressures, and very much tied up with what the corporations are doing.

Pavan Sukhdev says that the interplay between private sector activity and the interests of governments have become seamless: "Politicians are concerned about GDP growth, employment and fiscal deficit management, and doing all that in their present term of office. Then work out who is going to provide all this; it is the corporations. They are about 70 percent of the economy and 60 percent of employment and also the means to pay deficits with taxes. It's only natural, then, for the politicians to look over their shoulder to see if the policies they have are beneficial for the corporations."

And this is not the only big pressure for short-termism. In most democracies voters elect a new government every four or five years, so when we need to be looking at the future from the point of view of decades, we are constantly taken back into frames that look forward for at best a few years.

The result is that, although recent decades have seen modest regulation by governments to limit some of the worst damage that economic growth is causing to natural systems, there has been precious little effort in changing the economic system, so that it has a better fit with the Nature that keeps it going. On the contrary, in the wake of the recent financial crisis, some have gone into reverse. The 2011 statement by British Chancellor of the Exchequer George Osborne that "We're not going to save the planet by putting our country out of business" rather sums up how elected politicians relate to longer-term imperatives.

Remarks like this seem remarkably old-fashioned given what recent science has told us about the links between ecology and economy. As we've seen, 100 percent of the economy is dependent on Nature, the economy is degrading Nature, and this is translating into costs and risks for the economy. So at one level the reason for sustaining natural capital is about keeping the economy going, not Nature. So why don't economists get it?

Pavan Sukhdev says that in large part it's down to how they are trained: "A lot of the economists today have been schooled and got their degrees without ever looking beyond free markets and their rather artificial models. Crucially they have not been trained properly on the matter of externalities." He explains to me that an externality is the term used to describe a third party cost or gain that arises from a transaction. "For example, if I am a car manufacturer and I sell you a car, I am happy and so are you, but the lady across there breathes the fumes from the car that I made and which you bought, and she suffers health problems, then that is an externality, and in that case a cost for someone else. In Nature, the loss of pollinators, lost flood control, less wildlife, are all externalities of the economic system."

False Economy – or Bioeconomy?

From what I have seen through nearly three decades of ecological work, most of the economists advising world leaders appear to suffer from this kind of myopia. But if there were a stronger consensus among mainstream economists about the scale of the mismatch between the size of the growing global economy and the size and capacities of Nature, then what might the response look like? This is a huge question, and an important starting point in answering it is to look at the options that might be adopted. Essentially, there are three.

The first one is simple to explain: it is business as usual and seeking economic growth in the short-term, while drawing down more natural capital. This will lead to a continued disruption of the services and benefits provided by Nature, and will eventually lead to economic costs – for example, in raised insurance premiums, increased food and water prices and in the damage caused by climate change. And these costs could be catastrophic, as the chasm between our demands and what Nature can supply continues to grow.

In the debate on what to do about natural capital, it has become fashionable to declare that "business as usual is not an option." This sounds good, but is patently wrong, because it is exactly the option we have been pursuing. At best, it is softened with talk of "balancing" the needs of people and the environment, but the direction of travel remains broadly the same. It is not a good option, but it is on the table.

A second possible response is to innovate with technology – to keep the same basic economic system but to transform farming practices, water management, resource efficiency and our energy systems with technological solutions. In the process, the foundations for a super-efficient and more renewable economy would be laid. This sounds quite attractive, and technology will be a vital part of what is needed, but it won't be enough due to the sheer scale of the mismatch between the demands of economic growth and what natural services can provide.

On top of this it is important to remember that some of the things that Nature does for us cannot easily (or economically) be replaced by technology. Examples, as we have seen, include: the carbon storage functions of natural forests and soils; the productivity of the oceans; the work done by microorganisms in soils; the primary production carried out through photosynthesis; the protection of property by coral reefs; and the design solutions created by natural evolutionary processes.

While acknowledging important and powerful roles for technology, there is a third scenario. It might be summed up as the transition to a "bioeconomy." This word is already in use, but sometimes with a narrower meaning than what I am getting at here. The shift toward a bioeconomy could lead to the fusion between human economic development and Nature – biosphere and "buyosphere" seamlessly integrated, so that what we take from Nature doesn't diminish its ability to carry on providing what we need. Our economy would in effect become a part of Nature, rather than as now, where it is not only seen as a separate entity, but is actively engaged in a program of asset stripping natural systems, which in turn we celebrate as "growth."

The bioeconomy would be defined by its ability to support human needs indefinitely into the future, with the use of natural capital based on taking only dividends. It would encourage the careful management of natural systems, such that economic development strategies are conceived in tandem with a clear understanding of what Nature can and can't do, and where costs in terms of damage to natural systems never exceed the value of the benefits gained from keeping them intact. And if damage is caused to important systems, then there are mechanisms to compensate for the functions that have been lost.

In some ways the approach taken in this third scenario might parallel the philosophy behind *Biosphere 2,* whereby engineering and technology were deployed in the context of natural systems being maintained. Ethnosphere and biosphere maintained together, as though one was a subset of the other (which, of course, in reality is the case). And this fusion would operate not only in practical and mechanical ways, but in aesthetic, social and cultural senses, too. Markets could be a part of this bioeconomy, but only if the true costs of damage caused to Nature are reflected in the prices markets arrive at.

In the little bubble of *Biosphere 2,* eight people lived well by caring for a tiny piece of Nature, and it could work at the level of the Earth for 8 billion-plus people, should we choose to organize things like that. Biospherian Mark Nelson has something to say on this: "One of our tasks was to intervene when the natural diversity was threatened. The interventions were quite satisfying because it wasn't us and the environment it was us in the environment. We had a role in looking after it. Once we were in there we realized we were in the same lifeboat."

Placing our economies in Nature would require new institutions, laws, policies and culture. But could it be done?

A New Economics

Canary Wharf is home to one of the world's most important financial centers. Its iconic 1980s tower can be seen from miles around. As part of the City of London, the skyscrapers that have recently sprung up here seem as though they could not be further from Nature. But even this abstract world of stock prices, exotic financial instruments and reinsurance has roots that draw sustenance from the natural world.

There is a clue in the name, for this was the place where, in centuries gone by, bananas and other produce coming from the Canary Islands was once unloaded. The early wealth of this economic hub was thus founded on sunshine, volcanic soils and cloud forests.

And the wealth of the City still rests on these and other natural assets, although as time has passed the ever more extreme abstractions in our financial system have increasingly obscured that fact. But despite the conceptual distance we have placed between economics and ecology, it could still be possible to place Nature at the center of how we conduct development, and in the process to correct that debilitating short-termism.

I have spent quite a few years trying to convince companies that their commercial success will, in the future, depend on building intelligent relationships with Nature, and how this requires a sense of perspective beyond quarterly financial results. And it has been encouraging to see some positive responses, even the beginnings of a major shift.

There are quite a few places to glimpse this change. For example, in March 2012 I spoke at the Ecobuild trade fair in London. This annual event has grown very rapidly, from a few pioneers to a massive mainstream gathering with not only niche "green" enterprises showing off their wares but also major construction companies. Walking around the vast exhibition space at the Excel Center in East London, it is possible to see a new world emerging: 1,600 different exhibitors, between them presenting the solutions that could in a short time enable us to build and retrofit communities that are on their way to mimicking the ethnosphere built inside *Biosphere 2*.

As I browsed the exhibits, I thought about what John Allen had told me and about his philosophy in the design on that project. His idea was to create living space that nurtured the human spirit while at the same time protecting the biosphere that supports all life. Looking at the energy, commitment and inspiration on display, I was convinced that we could do this at a planetary scale, if we wanted to, to develop the ethnosphere in Biosphere 1 so that natural services are protected and sustained.

And it is not only at specialist trade shows that the possibility for this kind of shift is now apparent. Ecology minded leaders are emerging across a range of industrial and business sectors.

One that has led the pack is the American carpet company Interface. In 1994, the company's CEO, Ray Anderson, had what he described as a "spear in the chest" epiphany. It came from reading Paul Hawken's book, *The Ecology of Commerce*. Anderson was looking for ideas to include in a speech, but instead found inspiration that would set his company in a whole new direction. He called his idea "Mission Zero," and laid out the goal whereby the company would, by 2020, reduce its impact on Nature to zero.

To many eyes, the Mission Zero goal seemed an impossible dream. Anderson was undeterred, however, and got his team on the case, looking at the company from end to end. The result was remarkable. Not only did the environmental impact of the company begin to drop, but as it did so the business grew in size. New commercial models based on ecological drivers were found to have real value. One important change was to move out of carpet sales and into carpet leasing. This enabled new manufacturing and waste recycling technologies to be used, in the process increasing resource efficiency and cutting costs. The company is now also using plastic waste from the oceans to make some of its new carpets. This comes in the form of discarded nylon fishing nets collected from beaches and fed as raw material into the company's factories. Anderson himself died in August 2011 at age 77, but earlier that year he had judged that the firm had progressed beyond the halfway mark.

More recently, Anglo-Dutch consumer goods giant Unilever has emerged as a leader. In 2010, the company set out a Sustainable Living Plan so as to reach the goal of cutting its environmental impact in half by 2020. Unilever produces a wide range of brands, from Persil washing powder to Ben and Jerry's ice cream, Knorr soups and Lipton and PG Tips teas, and the company touches hundreds of millions of people every day. Its program has been backed by some bold signals from its CEO, Paul Polman.

When Polman launched the Sustainable Living Plan, there were questions from financial journalists who wanted to know what the new Nature friendly focus would mean for quarterly profit forecasts. Polman responded by saying that those investors who were interested only in short-term returns should take their money elsewhere. I'd never heard that from a major corporation before.

In one article, he wrote that the "short-termism" of so much modern business – "quarterly capitalism" – lies at the heart of many of today's problems and was quoted as saying that "if you buy into Unilever's long-term value creation model, which is equitable, which is shared, which is sustainable, then come and invest with us. If you don't buy into this, I respect you as a human being, but don't put your money in our company."

Retailer Marks and Spencer is also at the forefront, having set out its Plan A ("because there is no Plan B"), adopting ecological targets across its operations, from transport to waste and clothing to fish. Former Chief Executive Stuart Rose said after Plan A was launched that there was initially some internal resistance. Senior colleagues expected it would cost too much, confuse customers and alienate staff. In all three respects the opposite was in fact the result: it saved money, motivated staff and the customers welcomed it.

Another encouraging innovation has come from German sports company Puma, which set out a method of measuring its impacts in an Environmental Profit and Loss Account. This was introduced not only because of a desire to do the right thing, but because the company saw an opportunity to better manage some of the risks that confronted it, ranging from diminished natural resources to volatility in commodity prices and potential damage to its reputation.

Puma's chief executive Jochen Zeitz is one of the business leaders who has realized that a new context for commerce has emerged. "The business implications of failing to address Nature in decision making are clear," he says. "Since ecosystem services are vital to the performance of most companies, integrating the true cost for these services in the future could have significant impacts on corporate bottom lines."

And on this big point there has recently been a real and encouraging change of perspective in business. During 2011 and 2012, I worked with the University of Cambridge Program for Sustainability Leadership (CPSL) to help a group of private sector companies set out some clear views on what they should do about natural capital in advance of the 2012 Rio+20 Conference on Sustainable Development.

The statement they agreed on was signed by the chief executives of some of the world's large companies, and as well as pointing out the obvious facts – "the world has failed to respond to the challenge of sustainable development with adequate determination" – they went on to underline many of the challenges raised in this book. "Businesses' bottom line, and that of the entire global economy, is built on products and services provided by ecosystems and other components of natural capital," they declared. "Companies and governments must signal that the choice between economic development and sustaining natural capital is a false one, and take measures to create a global economy that pursues both goals simultaneously." It sounded like a small step toward the idea of a bioeconomy.

Among the companies that signed up to this compact was Kingfisher, the largest home improvement company outside the USA – and an enterprise whose demands on wood, board and paper requires a forest the size of Switzerland. Other signatories included Arup, Nestlé and Mars, each of whom is confronted with similarly compelling circumstances, in relation to water, soils and farm produce.

These attempts by companies to align their activities with Nature's capacities have in part been helped by consumer demand. For example, sales of certified "sustainable" timber products quadrupled between 2005 and 2007, while global sales of organic food reached about $46 billion in 2007, a three-fold increase on 1999.

While this is welcome, nobody has any illusions as to the motivation that most powerfully shapes business decisions – and that is, of course, profit. Where there is money to be made the temptation is always there to put Nature second, or third or, indeed, nowhere. This is why it will be essential to promote best practice by companies through transparent, robust and comprehensive reporting.

The manner in which companies presently tell the world how they are doing dates to Victorian times: company reports are required to focus only on financial information. Questions of natural capital are at best left to voluntary reports, which are often nowhere near as rigorous as the methods used for financial reporting. Nor are they comparable between different companies, and so of limited use to investors developing strategies based on natural capital priorities. This has prompted calls for "integrated reporting," providing far more extensive information, so that the opportunities and risks faced by companies, including in relation to natural capital, are properly set out. Many believe this needs to be required as a matter of law.

For the people in the financial sector who are moving that $80 trillion of investment around the world, such reports would provide clarity as to which companies are doing what, where there is risk in relation to impacts on Nature and where there is genuine leadership. If they had such information, they might finally be in a position to implement serious investment strategies geared up to the long-term protection of natural capital. And perhaps they will. Pavan Sukhdev believes "there is a serious global effort afoot to enable companies to calculate their externalities, and when you can calculate something, then your excuses for not managing it are gone."

But if private sector companies are to go the whole mile, and to do so fast enough, then it will be vital for governments to put in place the kinds of signals that will take them more rapidly in the right direction. A lot of regulation has been introduced in recent decades, to cut pollution and protect the most important natural areas, for example, but this won't be enough. If the benefits we get from Nature are to be maintained, then action is required not only in special places and in relation to particular pollutants, but right across national economies.

Mark Nelson, following his experiences in *Biosphere 2,* has reached this conclusion, too: "Conservation and preservation is all very well, but the real question is how to make humans economically viable without running the system down. Most of the world has done a good job in protecting the really fragile parts, the exceptionally important places, but the real issue is how do we live in this biosphere whereby we can benefit economically while keeping the environment intact?" There are no easy answers to that, but fortunately a few examples of how we might make a start.

Finding Nature's Value

In Chapter 2 we saw how Guyana's former President Jagdeo sought ways of realizing financial value from his country's extensive tropical rainforests. His plan was to realize some economic benefit based on the carbon held in the trees, and in so doing ensure that the services that the forests provide for the world were maintained.

Another developing nation that has taken countrywide action to find an economic value for Nature is Costa Rica. One of the leading figures in this has been the country's former energy and environment minister, Carlos Manuel Rodríguez. I have worked with Rodríguez on a series of programs to help the World Bank adopt more sustainable practices, and have been impressed by his journey and what he has achieved.

He explained to me how past deforestation in Costa Rica was driven by short-term economic pressures: "We are quite close to the USA and during the 1970s the fast food industry was growing rapidly. There was a huge demand for beef and we introduced large-scale ranching to meet it, and the rate of deforestation to make space for it was extremely high. During that decade we had the highest deforestation rate per capita in the world. At that time the market did not recognize the value of Nature and the forests."

National parks were set up in response, but this was not enough to conserve the country's huge variety of species. More was needed and, in the 1980s, attention turned more towards economic remedies. Rodríguez talks about how things began to change: "We understood that forest conservation can only be achievable if the value of the standing forest can match the cost of opportunity of not doing cattle ranching. So we put in place a lot of forest conservation incentives, using tools such as subsidies, tax breaks and direct payments. This began to lower the level of deforestation."

But it was not a politically sustainable strategy because there were big costs to the country's finances, and no clear way of measuring the benefits being gained: "The finance ministry guys couldn't properly measure the value of our investment. They were unable to see the value of keeping healthy ecosystems and the environmental services they were providing. We realized we needed to generate some information about the economic benefits of protecting Nature."

By the early 1990s Rodríguez and his team began to make a different case, at first using information about the value of Nature-based tourism: "Ecotourism was growing fast and we realized we could generate some information as to the value of protected areas." And then they realized it was possible to generate figures on the value of the forests to the hydroelectricity industry. The trees captured and released water and this then flowed in rivers that topped up reservoirs that powered turbines. Less forest meant less water, and that meant less power.

Having previously been told by the finance minister that health, education, new infrastructure and poverty reduction were priorities, and that Nature was not, Rodríguez presented his new information: "When we had this work completed I went back to the finance minister, but this time with some economists. When he saw these guys with me, he began to talk to them and they were speaking the same language. This was a turning point, and now the economics of Nature is institutionalized in Costa Rica." Not only were natural areas protected; large tracts of degraded land were restored, too. "In the late 1980s, we had only 21 percent of forest cover, but because of what we have achieved it is now 52 percent."

Rodríguez told me how this impressive outcome was achieved at the same time as improving living standards. "When the forest was at its low point the GDP was around $3,600 per person; now that the forest area has more than doubled, the GDP per person is around $9,000." And on the energy side too there was impressive progress: "In 1985, Costa Rica generated half of its energy from renewables, and half from fossil fuels. Twenty-five years later we generate 92 percent from renewables."

Rodríguez is quick to point out that this doesn't mean that if forest cover is restored and the energy sector goes green, then per capita income necessarily goes up. What he says it does show, however, is that natural capital can be protected and restored at the same time as achieving economic development. This is really important to highlight, as it is something that many governments evidently still do not understand or believe.

And in Costa Rica, the valuing of Nature has led to a whole new economic and political mind-set. Rodríguez says: "We have realized that there is no long-term economic growth without protecting the health of the ecosystems. Every year we carry out an ecological footprint, and with the economic and social information included, so as to find out how we are doing as a country. If we can show that economic and social health is dependent on the health of Nature, then most politicians see the case we are making. This is all about humans, not about Nature." This is a kind of integrated reporting, applied to a nation state, and the power of that is clear in the outcomes seen in this remarkable country.

The inspiration offered by Costa Rica is all the more powerful because it is a developing country. I asked Pavan Sukhdev where he saw leadership among the more developed nations. "Number one I point to the Norwegians," he said. "They have used their oil money for the public good, for example, in pledging $1 billion each to Indonesia and Brazil to help cut their rate of deforestation. They have this money and use it for the public good, doing what they can to correct externalities."

It would be hard to state a place for my country, the UK, in the same leadership category, with the backward looking views recently expressed by its leaders, but the country made an important step in 2011 with the publication of its first National Ecosystem Assessment. This impressive piece of government-backed research reached similar conclusions to many of those set out in this book. It pointed out how Nature provides a wide range of benefits vital for the UK, underlined their huge economic value, and noted that some services are in decline and that new ways are needed to retain them. Its detail identified a wide range of values attributable to the UK's ecosystems. For example, it valued the carbon taken up by UK woodlands at about £690 million per year. The benefits derived from improved river water quality were found to be about £1.1 billion per year. The value of coastal protection provided by wetlands was estimated to be about

£1.5 billion per year. The amenity value of inland wetlands added a further £1.3 billion per year.

During 2012, I was involved in projects to advise the UK government on how best to translate this assessment into mechanisms that would enable companies to develop business strategies that value and protect natural capital. While there is a lot of complexity, there is no show-stopping reason as to why this can't be made to work. From changes to the tax system to different indicators of economic performance, and from trading schemes to new priorities in land use planning, governments have a range of options.

One approach that is attracting growing international interest is "biodiversity offset." These mechanisms are founded on the idea of achieving "no net loss" of Nature, such that damage in one place is compensated for by protection or restoration elsewhere. At the time of writing, there are thirty-nine national compensation programs geared towards the idea of no net loss, with twenty-five more in development. They range hugely in size and ambition. In the USA, for example, there is a national "wetland banking" program which offers incentives to create new wetlands to compensate for those that are drained or built upon. In Australia, each state has a no net loss scheme in operation.

It is wise to be cautious about no net loss schemes. Some create questionable benefits and, when Nature is traded in a market, there can be unintended outcomes. However, if used as a last resort after other alternatives have been eliminated, properly regulated and with effective transparency, such schemes can have an important and positive role. Richard Burrett is among those who see real potential in economic mechanisms: "There are markets, for example, in wetland offsets, so that when one is damaged another is created and people can buy wetland credits if a development damages an existing one. This creates an incentive for organizations to rehabilitate wetlands so that credits can be sold to developers. It's already a $2-3 billion market. The same thing can be done for nutrients."

These economic mechanisms can perhaps best be seen as a bridge across the chasm between our short-termism and the long-term benefits that Nature supplies. There are plenty of tools available and the challenge is for these to be used at sufficient scale, which will require leadership and vision.

Economism and Religion

To go further and faster, we mustn't lose sight of our cultural attitudes towards Nature and recognize that the discourse goes far beyond a debate over particular economic tools.

Take, for example, the value we attach to carbon. For what are evidently reasons of culture, we treat the carbon in trees as for the most part worthless, whereas diamonds, made of pure carbon, are exchanged for huge sums. Diamonds have some industrial applications but, considering the climatic benefits of carbon in trees, this comparative valuation is utterly perverse. Economic signals are often more of a social construct than any rational assessment of where actual value lies.

During my visit to the páramo above Bogotá in Colombia, for example, to see the scheme to protect its water resources, I was told how the better security situation in that country was attracting keen interest from international mining companies, including those interested in prospecting for gold. Because of the recent financial crisis, the value of gold had soared and for the mining companies a potential killing could be made. But, while the yellow metal inspires fortunes to be invested for its extraction, the water and forest systems that will be impacted by mining still struggle to find the kind of economic status that will keep them intact, despite their fundamental importance for humankind through the many services they provide.

With such examples in mind, and the (irrational) refusal of many mainstream economists to take proper account of natural capital, some argue that economics has taken on values to the point of being a religion – "economism." Pavan Sukhdev has thought about this a lot: "There is an unstated religion in economics, to the point where it is believed that everything can be resolved with free markets. The ghost of neoclassical economics and a few leading thinkers in that field continue to exert their influence on generations of young economists who go to work in national treasuries and who don't understand what natural capital is all about. The question is, how do we address this legacy?"

A good question – and one that begs another, which is about where our values come from, especially those that have the potential to generate longer-term perspectives.

Like the skyscrapers of the financial centers, the Basilica of Saint Francis in Assisi can be seen from miles around. It is a symbolic embodiment of values once as powerful as those that drive the modern City of London. And it has a very particular status, for it carries the name of a man who became known in the Christian faith for his close affinity to the natural world. Giotto painted frescoes in the Basilica of Saint Francis preaching to the birds. These famous images remind us of his spiritual connection with the natural world, and his belief that creation was a sacred gift from God.

This outlook was, of course, at the heart of Christianity. People were believed to be embedded in creation, and therefore in the presence of the Almighty. But times changed, and Christianity and other major faiths have drifted somewhat from their original perspective and now embrace a world view that sees Nature as a collection of resources placed on Earth for the material betterment of humankind.

I was in Assisi in November 2011 to mark the launch of a Green Pilgrimage Network set up to promote awareness among the tens of millions of the faithful who annually embark on pilgrimages to sacred sites, from Mecca to Amritsar, Louguan to Santiago de Compostela. The network encompasses fifteen faith traditions and is part of a longer-term initiative coordinated by the Alliance of Religions and Conservation to put Nature back at the heart of their message.

Although I follow no religious faith, I was invited to speak at this unique multi-faith gathering and raised the ways in which we should place more value on Nature, economically and morally, and how religions could renew spiritual relationships with the natural world. If this were to occur, I argued, it would be more likely that economics could follow a different path.

This is dangerous territory, of course. Great biologists like Richard Dawkins and Edward Wilson are among those who question the role of faith as a way to embrace ecological and other challenges. I understand their point, but beg to differ, for the unfortunate fact is that the majority of humankind either doesn't care about ecological science or doesn't respond to the messages it brings. Meantime, some 80 percent of the world's population is aligned with one religion or other. If the values promoted by the faiths can change culture in ways that will help us realign our demands on Nature, then they are part of the solution.

The world's great faiths are not the only potential source of philosophical inspiration for new ways of making sure we keep Nature in good shape. Another potential source of change to how we collectively look at the world could come from one of the world's largest industries — advertising. Employing some of the world's smartest psychologists and communications experts, this industry is responsible for a lot of our behavior and much of our world view. Might it be possible for this powerful industry to be part of the solution?

Again, I think it has to be. No amount of hand wringing as to the evils of advertising is going to change how it works, but some intelligent engagement with its practitioners might just lead to harnessing some of its massive power in the pursuit of different ends.

Gardening the Earth

Despite our incredible technological capabilities, humans have managed to travel from Earth no further than our nearby Moon, under 1.5 light seconds away. Able to survive for only a brief period without the Earth's natural systems, people have been forced to come back to Biosphere 1. The few space stations that have maintained people aloft for longer periods have had to be constantly resupplied from natural capital here on Earth – with food, fuel and other essentials.

Even unmanned vehicles are as yet to leave the solar system. The two Voyager spacecraft launched in 1977 set out to explore the outer planets and then to embark on journeys into interstellar space. Should either one of these craft be found by intelligent life, they carry gold-plated audiovisual discs with our story, images of people and the wonders of the Earth. After thirty-five years, they have reached the edge of our solar system – and that is about the extent of the experiment. The next nearest star to the Earth after our Sun is Proxima Centauri, a pinprick of light about 4.2 light years away. Travelling at 60,000 kilometers per hour, it would take a Voyager craft about 76,000 years to get there. But even if one of these craft did get there, orbiting that star are no earthlike planets that could support life.

The Voyager mission is a stark reminder of how Biosphere 1 is indeed our only home. There is nowhere else to go. No matter how clever our financial system, impressive our rates of economic growth or sophisticated our technology, there is no place to move to should we degrade our biosphere to the point where it can no longer meet our needs and sustain our economies.

In ecological terms, the coming decades are set to be the most momentous for millions of years. The good news is that we can anticipate rising human demands on Nature and manage them with a wide range of tools. Many are already in use, and their effectiveness already demonstrated, while others are in development; the challenge is to refine them and bring them to scale. Crucially this will rely on changes to economics and, equally crucially, on the popular culture and philosophical outlook of societies that shape the choices we collectively make.

Perhaps it would help if we begin to see Nature for what, at one level, it so obviously is – the source of essential services: a provider of insurance, a controller of disease, a waste recycler, an essential part of health provision, a water utility, a controller of pests, a massive carbon capture and storage system and as the ultimate converter of solar energy.

Looking forward, is there really any debate as to the extent that we will need Nature to provide all this? And, indeed, to need it more than now, with our rising global population? But all too many of the people who run our world – finance ministries, presidents, bankers, the CEOs of global corporations – behave as if this is some kind of mythology, not real economics at all, and a mere marginal question. Better, they argue, to promote growth and development and our problems will be solved.

Despite such myopia, there is a growing realization of the vital economic value of Nature.

But can Nature continue to deliver as the human population edges up towards the 9 billion-plus people projected by 2050?

Many analysts believe that it can – and that our biosphere, if treated properly, can provide economically vital services indefinitely. There is a caveat, of course: for 9 billion people to live in harmony with the Earth, life will be rather different to today.

Perhaps as we move from 7 to at least 9 billion we can get a few clues from imagining what it might have been like had there been an attempt to increase the number of people living in the microcosm of *Biosphere 2*. Might the elevated demand have required some changes to behavior, since the biosphere remained the same size? Might even more care have been needed to maintain the systems that were meeting people's needs? The answers go without saying. The only difference between the inevitable impacts on the people in the real world of Biosphere 1 is the time scale.

The simple conclusion I reach is that we need to take a different approach to how we look at Nature and the Earth. If we can do that, then Nature can be maintained and enhanced, for the benefit of people and the rest of life, indefinitely into the future. We need to garden the Earth, to nurture and husband its assets, aware of the implications of our decisions. We need to produce food and develop cities in ways that keep natural systems intact and capable of discharging their essential functions.

Key to making this happen is the realization that Nature is not separate from the economy, a drag on growth or an expensive distraction. We know all we need to do things differently. Biosphere 1 still works and we can keep it that way, if we wish to.

The alternative is to carry on as we are now. After all, what has Nature ever done for us?

ACKNOWLEDGEMENTS

IF I HAVE MADE ANY PROGRESS in producing an accessible and readable account of what Nature does for us, it is in large part because of the willing assistance I have received from so many friends, colleagues and other experts who shared their knowledge and ideas, helped with research and commented on drafts.

My friend Heather Buttivant helped me with extensive research across much of the title. I am indebted to John Allen, Mark Nelson, Deborah Snyder, Chili Hawes and William Dempster, who all helped me by sharing their experiences of Biosphere 2. Jane Rickson and Jim Harris at Cranfield University told me about their work on soils. Martina Girvan shared her work on soil microorganisms, and Joe Bull (among other things) his work on saiga antelope. I am grateful to Paul McMahon, who helped me in explaining how grazing animals can increase soil carbon and to Alan Knight for giving me further views on this subject. Philine zu Ermgassen at the University of Cambridge gave me invaluable advice on the services provided by oyster reefs.

Andreanne Grimard at the Prince's Charities International Sustainability Unit helped me on issues linked to forests and carbon storage, as did Kevin Hogan from the office of the President of Guyana. I am grateful to the South Australia Tourism Commission for their assistance in helping me visit very special places on Kangaroo Island and in the Flinders Ranges in the Outback of South Australia. I am most grateful to Ran Levy for his efforts in arranging such an informative trip in Israel, and to David Furth and Ruth Yahel, whom he arranged for me to meet and who provided me with very valuable insights.

My daughter Maddie Juniper did the research and number-crunching that made it possible for me to describe my train journey as presented in Chapter 3. Alison Austin from the Robertsbridge Group sent me background material and provided essential encouragement. Robin Dean was generous with his time and advice on pollination, as was William Miller on New Zealand bees.

Chris Bowden and Paul Morling at the RSPB and Stuart Butchart of Birdlife International all helped me with advice and research on how wildlife helps to control pests and diseases. José Yunis at the Nature Conservancy, Juliana Ocampo Herrán at Bavaria in Bogotá, and Carlos Florez, Manuel Rodríguez and Mathieu Lacoste all helped make a trip to Colombia both informative and enjoyable. Professor Neil Burgess helped me better understand the importance of cloud forests and provided very helpful comments on my draft manuscript.

Alan Rodger at the British Antarctic Survey gave me valuable advice, as did Dr. Carol Turley at the Plymouth Marine Laboratory. Professor Callum Roberts at York University was a source of helpful advice on marine ecosystems and I was very pleased to be able to speak with Jo Royle about her experiences on the high seas. I am grateful to Hai Nguyen, who translated conversations with Vietnamese fishermen and to Lucy Holmes for being such a great travelling companion while investigating fisheries management in that country. Sebastian Troëng at Conservation International in Washington DC also helped me navigate aspects of the oceans chapter. Sonia Roschnik at the National Health Service Sustainability Unit in Cambridge gave me invaluable advice, as did Dr. William Bird on his experiences in looking at how Nature nurtures human health.

Richard Burrett of the United Nations Finance Initiative helped me, as did Pavan Sukhdev and Joshua Bishop of the TEEB process, in explaining their ideas on where ecology meets economics. I am also grateful to Carlos Manuel Rodríguez for the help he gave me in this respect and to Jack Gibbs for his views on the financial system.

I received very helpful general advice from Mark Everard, who among many other things works with the Environment Agency of England and Wales, and from Laura Somerville and her colleagues Esther Bertram, Ros Aveling, Erin Parham, Elizabeth Allen, Tony Whitten, Alex Diment, Rob Brett and Mark Infield at Fauna & Flora International, who similarly helped me by sharing their extensive knowledge of the subjects covered in this book, including, for example, the impacts of sea otters in kelp forests. I am especially grateful to Laura Somerville, who also helped me with picture research.

Johannes Förster at the Department of Computational Landscape Ecology in Leipzig pointed me towards much helpful research. My friends David Edwards, David Barley, Charlotte Cawthorne, Edward Davey and Klaas de Vos at the Prince's Charities International Sustainability Unit all helped me in different ways. Katie Hill, senior researcher and writer at *Which?*, offered me very helpful reactions to the draft, additional focus to the narrative and invaluable encouragement in helping me see the value of this project. My friend Louise Bell was kind enough to read an advanced draft and to give me very helpful reactions to it.

I am grateful to Mark Ellingham at Profile Books, who saw potential in the title, commissioned it and provided much invaluable editorial advice. Warm thanks also go to Ruth Killick at Profile Books for her great work in raising awareness about the title, and to Nikky Twyman for her expert proofreading.

I am also immensely grateful for the support I received from Verity White, Sue Gibson and Ellie Williams in making a short film on public reactions to the question posed by the title of this book (this can be found at YouTube via the title of the book). My wife Sue Sparkes provided invaluable support and I thank our children, Maddie, Nye and Sam for their patience while this this book was being written.

I have tried hard to avoid errors, but if there are any, they are mine alone.

For the production of the handsome U.S. editions of *What Has Nature Ever Done for Us?* I am immensely grateful to Deborah Parrish Snyder and her team at Synergetic Press: Linda Sperling, David Rogers, John Cole, Jim Mafchir, Debbie McFarland, Graciela Ruiz, Jeanie Williams, Kayle Schnell and Mitch Mignano, for making this book such a marvelous product for North American readers. Sue Sparkes and I also appreciate the time and effort Deborah and her team put into arranging a wonderful U.S. promotional tour for the book in the summer of 2013; including a very memorable stay at Synergia Ranch in New Mexico, where it was our great pleasure to meet several key figures of the Biosphere 2 experiment. I am also very pleased to acknowledge Phil Siegel from Media Works in San Francisco, Jackie and Jeannette from Conscious Media Relations in Los Angeles and Jen Bergmann in Seattle, who all made invaluable contributions in raising awareness about the title.

Tony Juniper, Cambridge, 2013

Index

PHOTO CREDITS

ABOUT THE AUTHOR

Tony Juniper is a leading independent environmentalist whose many activities include being Special Advisor to the Prince of Wales Charities' International Sustainability Unit; Fellow of the University of Cambridge Program for Sustainability Leadership and a founding member of the Robertsbridge Group, which advises international companies. He is the author of the award winning *Parrots of the World*; *Spix's Macaw*; and *How Many Light Bulbs Does It Take To Change A Planet?* He also coauthored *Harmony*, with HRH The Prince of Wales and Ian Skelly. Juniper began as an ornithologist with Birdlife International. In 1990 he joined Friends of the Earth, becoming Executive Director (2003-2008) and serving as Vice Chair of Friends of the Earth International from 2000-2008. He has chaired the Advisory Board of the UK Action for Renewables campaign and served as Editor-in-Chief of National Geographic *Green* magazine.

"Juniper explains how the welfare of the human species rests on the assets and services provided by the rest of nature, and makes the case for natural capital to be integral in a new economy fit for the future."

"Tony Juniper takes us on a highly readable, personal journey of discovery of nature and our reliance upon it. *What Has Nature Ever Done For Us?* provides the stories and the numbers to convince others that investing in nature's balance sheet is good for the corporate balance sheet."

Juniper lays out the many ways that natural ecosystems pay dividends to human societies. ... This is important stuff that policymakers and the public need to be reminded of.

Bob Holmes,
New Scientist

"*What Has Nature Ever Done For Us?*
is a brilliant resumé of nature's new deal: nurture me and I'll nurture you."

Nick Crane, author and TV presenter